FIRST STEPS IN
ENAMELING

JINKS M^cGRATH

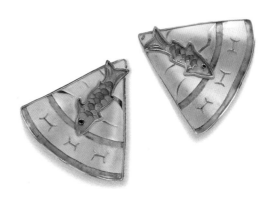

FIRST STEPS IN
ENAMELING

JINKS M^cGRATH

THE WELLFLEET PRESS
WELLFLEET

For Steve
an uncompromised craftsman

The author and publishers would like to thank
Gudde Jane Skyrne of Camden Workshops in
London for invaluable advice and information,
and for compiling the list of suppliers on page
96; and the British Society of Enamellers for
generous use of their photographic resources.

A QUINTET BOOK

Published by The Wellfleet Press
A Division of Book Sales, Inc.
114 Northfield Avenue
Edison, New Jersey 08837

This edition produced for sale in the U.S.A.,
its territories and dependencies only.

ISBN 0–7858–0033–6

This book was designed and produced by
Quintet Publishing Limited
6 Blundell Street
London N7 9BH

Creative Director: Richard Dewing
Designer: Ian Hunt
Senior Editor: Laura Sandelson
Editor: Lydia Darbyshire
Photographers: Jeremy Thomas, Nick Bailey

Typeset in Great Britain by
Central Southern Typesetters, Eastbourne
Manufactured in Singapore
by Eray Scan Pte Ltd.
Printed in Singapore
by Star Standard Industries Pte Ltd.

PUBLISHER'S NOTE

**Metalwork and enameling can be
dangerous.** Heated kilns and their contents
can burn, and acids can destroy and maim
if appropriate safety precautions are not
taken. Some of the chemicals used in
enameling are poisonous, and some
contain lead. Lead-free enamels are
available.

Always follow the safety instructions given
in this book, and on packages and labels,
when enameling. Always read the
instruction manuals provided by
manufacturers of kilns and other
equipment. Follow manufacturers'
instructions concerning the installation and
use of electrical and gas-powered devices.

Wear protective gloves and other clothing
when enameling. For clarity, some of the
photographs in this book show technical
processes being carried out without the
protection of gloves. This is not
recommended – never use a kiln without
wearing heat-resistant gloves; plastic
gloves are a recommended safety measure
when working with acids.

All statements, information, and advice
given in this book regarding methods and
techniques are believed to be true and
accurate. Neither the author, copyright
holders, nor the publisher can accept any
legal liability for errors or omissions.

Contents

Introduction

My interest in enameling began when I was studying jewelry-making. My tutor was an enameler, and although I never saw him at work, some photographs of gold enameled boxes he had made years before inspired me to make enameling the most important part of my course. After I left college I did not do much enameling because I did not have a kiln or a suitable working area, but my interest in the subject was reawakened when I went to a wonderful exhibition of René Lalique's work at Goldsmiths' Hall in London in 1987. I just wanted to hold and discover the secrets of all those wonderful pieces.

As is the way of things, enameling at first glance seems to be a comparatively small subject, but it mysteriously unfolds as you start to appreciate just how many uses there are for glass fused to metal for decorative purposes – old-fashioned badges, for clubs, team captains, and prize-givings; coffee pots and mugs; dog bowls; old glasses and decanters; modern pens and pencils; inexpensive jewelry; old-fashioned advertisements, railway signs, and garage signs; silver teapots; magnificent old gold jewelry; iconic panels; coats-of-arms. All these items and many more come alive as you start to notice and appreciate the art of enameling. Until we are involved in the craft, it is so easy to take these things for granted, little realizing just what goes into achieving that fusion of two media, and how it is planned, designed, and patiently worked on for hour after hour.

I hope that this book will inspire you to explore the possibilities of

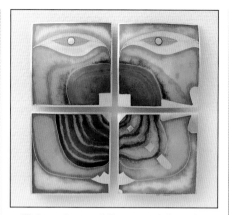

Cloisonné enamel jigsaw mask brooch.
GUDDE JANE SKYRME.

enameling, and, by including photographs of some of the work of Britain's best enamelers, I hope it will give you some idea of what can be achieved. As long as you follow a few basic rules, you should be able to achieve good results in a short time, even if you are a newcomer to enameling. When you have mastered the early skills, experiment with the materials to see what works and what does not. Try to work out why something turns out as it does. Enameling is a very individual craft, and you will soon develop a way and style of doing things to suit yourself. Different people have different ideas about the right way of doing something, so do not be afraid of doing it the way it works for you. If you have a problem, discuss it with another enameler if you can. Be logical about working things backwards when you are trying to solve a problem. Read other books on the subject, or attend a class. The techniques and methods described in this book really relate to my own experiences with enameling, but I have discussed ideas and experiences with other enamelers, and the guidelines I give don't *only* work for me.

The ground rules of enameling are easy to define: your working area must be perfectly clean, and you must prepare your metals and enamels in the correct way. The success of a finished piece largely depends on how carefully these rules have been observed. Beyond that, however, are so many subtleties and individual preferences and ways of working that I can only suggest, guide, and, I hope, steer you in the right direction.

Any book of this nature that does not include a section on how to deal with the unexpected is, I think, being rather unkind to its readers, so I have included some ideas and suggestions to help you deal with the inevitable "What do I do now?" moments. I'm sure that I have not covered every eventuality, but I have found that being brave sometimes helps! I think of what might help me out of a problem, take the bull by the horns, and, nine times out of ten, it works. There will, of course, always be that tenth occasion . . .

If there is one thing I have learned about enameling since Lalique cast his spell at Goldsmiths' Hall, it is that it does not take long to become obsessed by its charms and willful ways. Once you are obsessed you will find that you will be constantly seeking for that elusive "perfect piece." How often have I spent hours and hours on a piece, only to find some little flaw, a color that is not quite right, a tiny aspect that I wished I had done differently, and said to myself that the next time, it would be different – that one I would get right.

Let me introduce you to that world. Make of it what you will!

1
Equipment
And
Materials

THE KILN

Apart from a large piece of silver or gold, a kiln is the most expensive item you will need, and although it is possible to do some enameling very successfully without one by using a blow torch instead, at some stage a kiln becomes essential. You can choose between gas-fired and electric models, both of which are available in a variety of sizes, so try to make sure that when you choose a kiln the firing chamber will suit the kind of work you intend doing. For example, if your interest lies only in small pieces of jewelry, a chamber about $4\frac{1}{2} \times 4\frac{1}{2} \times 2\frac{1}{2}$ inches may be large enough. However, if you intend enameling larger articles, such as bowls, plates, and tiles, you should look for a kiln with a firing chamber that measures $9 \times 9 \times 6\frac{1}{2}$ inches. Generally speaking, I would choose the largest kiln I could afford and had space to accommodate.

Before you buy, remember that there are dozens of secondhand kilns waiting to be rediscovered. They turn up in dusty corners of school artrooms, attics and workshops, and I have been lucky enough to have had two kilns passed on to me that were no longer being used by their owners. It is worth telling your friends and acquaintances that you are on the look out for one, and the "for sale" and "wanted" advertisements in local papers or craft magazines may also unearth a redundant kiln, so try these avenues before assuming you must buy a new one.

Electric kilns

An electric kiln can be fitted anywhere near a suitable socket. Most enameling kilns will run off the domestic electrical supply, but if you are buying a larger kiln check this first. It should stand on a heat-resistant surface, in a draft-free area. Electric kilns have elements

Regulator with a 1–10 dial, which has a socket plug and electrical cord that connects with the kiln.

inside their walls, and it will take at least an hour for an average-sized kiln to reach a firing temperature of between 1,475–1,800°F. To help gauge temperature, a kiln can be fitted with a regulator or a pyrometer. A regulator is a "Simmerstat" with a 1–10 dial. It is plugged into the kiln and set to stop the kiln heating past a certain temperature. A pyrometer is more expensive than a regulator. It is a probe, which fits into the back of the kiln and is connected to a temperature gauge. It reads the exact temperature in the kiln, and so allows you to be sure of the temperature before firing.

If it is used correctly, an electric kiln will last many years. It is possible to replace blown elements, but, unless you can fit the element yourself, the cost of doing this probably means that it is time to get another kiln. Before you switch on a kiln, it is worth wiping out the inside chamber with a damp cloth to remove any bits of firescale or other hazards that may have accumulated during the last firing. It is a good idea to keep a ceramic mat at the bottom of the kiln, and this can be replaced by another mat when it gets too dirty or covered in burnt-on enamel.

Electric kiln with pyrometer probe which fits into the back of the chamber.

Gas-fired kilns

These kilns, which run off either bottled propane gas or natural gas, are available in a wide range of sizes. They heat up more quickly than electric kilns, and they are, therefore, able to make up heat lost through an open door more quickly during firing. Metal is slightly less prone to oxidization in gas-fired kilns, which have a reducing, rather than a reflective, heat.

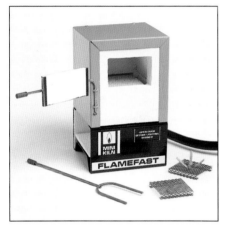

Mini gas-fired kiln, which is useful for small pieces as it heats up very quickly.

If your kiln does not have a regulator or temperature control, turn it off or leave the door open for a few minutes every hour or so if it is becoming a very bright yellow. The optimum working color is a bright red/orange, and continuous heating much over this will shorten the life of your kiln.

Following is a guide to the colors and approximate temperatures of your kiln. You may find you never need a pyrometer if you know the colors in your kiln before firing. The colors indicate similar temperatures in electric and gas-fired kilns.

Dull red	1,350°F+
Dull, orangey red	1,400°F+
Cherry red	1,450°F+
Bright red/orange	1,500°F+
Very bright orange	1,600°F up to 1,870°F
Very bright yellow	1,870°F+

If you are enameling without a kiln, support your work on a wire mesh on a tripod, at a height that allows you to play your flame from underneath. Remember that the enamel should not come into direct contact with the flame, so place your work with the enamel side up and, if possible, arrange some fire bricks around the tripod to give some reflected heat to your work. Try melting enamels on test pieces first so that you will have some idea of the likely results.

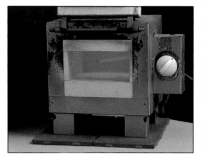

Dull red

Dull, orangey red

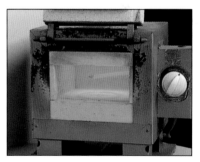

Cherry-red

Bright red/orange

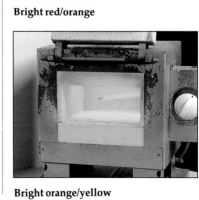

Bright orange/yellow

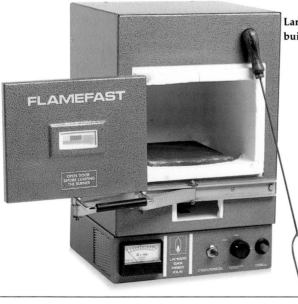

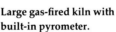
Large gas-fired kiln with built-in pyrometer.

TOOLS FOR ENAMELING

The tools you need are not expensive. You can make some of them yourself, and you may be able to acquire all sorts of useful stainless steel picks, spatulas, and so on from your dentist. Save glass and plastic containers with lids from the kitchen.

You do not need a large area to work in, but the most important thing for an enameler is to be able to work in a clean space. Choose a draft-free, fairly quiet area. The fewer distractions there are, the better, for once you have placed a piece of enameling in the kiln to be fired, it needs your undivided attention. Everyone has tales to tell of how, when the telephone rang, a piece that had taken hours to make finished up unrecognizable.

Listed below are a selection of the tools you need for basic enameling. There is no need to get them all at once. You need the 20 basic tools to start with, but you can collect the others.

Basic tools

Cooling tile Place hot pieces on a tile when you take them from the kiln.

Dental and homemade tools These tools are used to place the wet enamel on the metal.

Duckbill metal cutters These large, metal cutters are sometimes held in a vice when cutting.

Glassfiber brush This is used to clean the metal before enameling.

Glass screw-top jars Keep a selection of jars for washing powdered enamels.

Graded sifters These are used with dry enamels.

Gum The surface of the metal is given a coat of gum to hold the enamels.

Heat-resistant gloves Always wear these when you are opening and closing the kiln door.

Long firing fork You will use a fork to lift mesh stands into and out of the kiln.

Metal cutters These are used for cutting metal to size.

Mica A piece of mica is useful for placing enamels on when you are firing; it is also used for plique-à-jour work.

Mortar and pestle These are used to grind lump enamels.

Piercing saw A handsaw with very fine blades is needed to cut metal accurately.

Plastic containers with mesh (screw-top) lid Use these containers when you sieve dry enamels.

Quill A quill is used to place wet enamel on the metal.

Sable paintbrushes You need a variety of quality paintbrushes.

Small hand drill You will need a small drill to make holes for fittings and jewelry findings.

Stainless steel mesh stands Place your work on a mesh stand before putting it into the kiln.

Stainless steel stands or trivets Use a stand to support the work.

Wet and dry papers You will need grades 240, 400, 600, and 1,200 for cleaning the surface of the metal.

Supplementary tools

Asphaltum Paint the compound on metal to protect it from acids or etching solutions.

Burnishers These are used to burnish and clean edges.

Carborundum stones These are used to stone down enamels.

Cloisonné wire Reels of fine silver, copper, or gold round wire, in different diameters, 0.2–0.5mm, are used in cloisonné work.

Engravers These prepared steel tools are used for engraving designs on metal, and are usually associated with basse-taille work.

Ferric chloride A much slower etching medium for copper is made from 7½ parts water to one part ferric chloride.

Files Use a selection of files to clean up the edges of pieces of metal.

Formers Wooden or steel domes and a doming block are used to shape metal.

Nitric acid A solution in the proportions of between three parts water to one part acid and eight parts water to one part acid is used to etch silver and copper. As with any acid, **always add acid to water.**

Palette You will need a ceramic dish with approximately 10 sections to hold wet enamels while they are being used.

Palette knife A flexible, metal, spatula-type knife is used for mixing painting enamels.

Pumice powder The powder is mixed with water to form a paste that is used with a toothbrush to clean metal.

Rawhide mallet This mallet is used for shaping metal without leaving marks.

Sandbag A sandbag creates a useful soft surface on which to shape metal.

Sulfuric acid A solution in the proportions of one part acid to 10 parts water is used for pickling silver, copper, and gold. **ALWAYS add acid to water.** Never pour acid into a pickling jar before you add the water; always put the water in the jar, then add the acid.

Turpentine Use turpentine to remove Asphaltum from etched pieces.

Tweezers These are used for handling cloisonné wires.

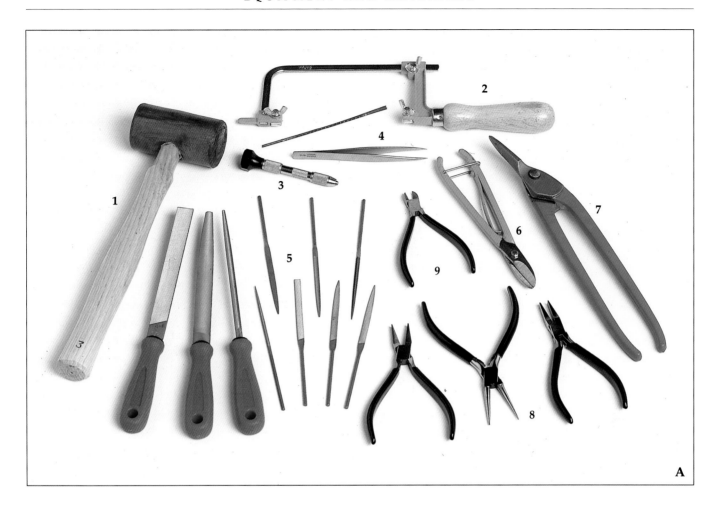

A

B

KEY A

1. Rawhide mallet
2. Piercing saw and blades
3. Small hand drill
4. Tweezers
5. Selection of files
6. Metal cutters
7. Duckbill metal cutters
8. Round and flat-nosed pliers
9. Side cutters

KEY B

1. Mortar and pestle
2. Graded sifters
3. Selection of dental and homemade tools
4. Quill
5. Sable paintbrushes
6. Plastic containers with mesh (screw-top) lid
7. Palette knife

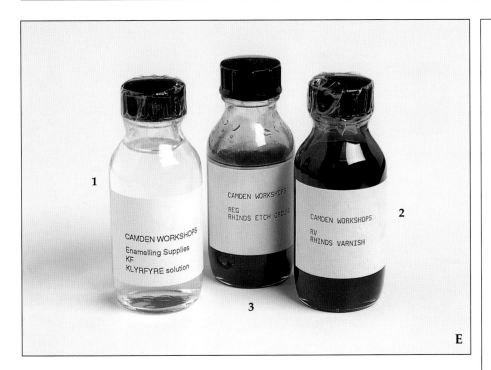

E

MAKING YOUR OWN TOOLS

You can easily make a spatula, pointer, and spreader for yourself.

You will need
3 lengths ¼ inch wooden dowel, each about 4 inches long.
Small drill
3 metal paperclips
Hammer
Small flat file

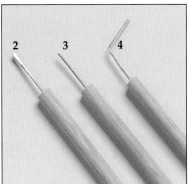

KEY C

1. Pumice powder
2. Glass brushes
3. Carborundum stones
4. Burnishers
5. Wet and dry papers

KEY D

1. Mica
2. Stands or trivets
3. Mesh stands

KEY E
1. Gum
2. Asphaltum
3. Etching ground

KEY F
1. Heat-resistant gloves
2. Long firing fork

1 Drill a small hole up one end of each piece of dowel.

2 Straighten a paperclip, and use a hammer to flatten one end. File it so that the end is a smooth oval. Snip off the excess length, and insert it into the dowel. This spatula is like a little spoon, and can be used to pick up enamel out of water and place it onto your work.

3 Make a pointer by using a file to sharpen one end of a straightened paperclip. Snip off the excess length and insert it into dowel. This tool is used to push wet enamel down and into difficult corners.

4 Make a spreader by bending the end of a straightened paperclip to an angle of about 45 degrees, snip off the excess length, and place it in dowel. Use this tool for spreading the enamel evenly over the surface of the metal.

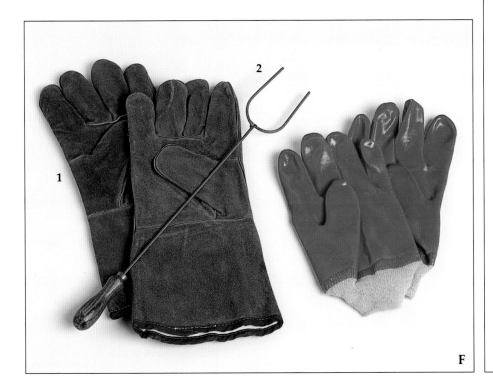

F

MAKING A SIFTER

1

6

1 Find a glass jar with a screw-on lid, and scribe a circle about half the diameter of the lid in the center of the lid.

2 Drill a hole on the line, insert a saw blade or small penknife blade, and cut away the inside of the lid.

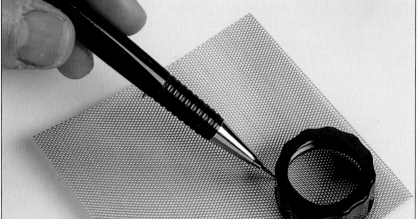

3

6 Place the mesh circle and the cardboard washer into the cut-out lid, pushing them in so that they fit snugly.

7 Screw the lid back on the bottle.

3 Use a pencil to draw around the base of the lid on a sheet of stainless steel or bronze metal gauze, approximately 80 mesh.

4 Cut out the mesh circle with a pair of scissors.

5 Cut out a cardboard "washer" the same diameter as the mesh circle, and then cut out the inner circle of the washer.

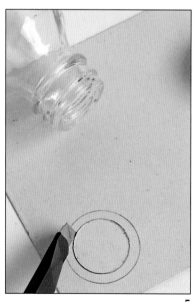

5

7

Enamels are a mixture of silica (alkali compounds which lower the melting temperature of the silica), lead oxides, salts of soda, potassium and boric oxides. Colorless enamels are known as fluxes, and the colors are obtained by adding various metallic oxides and/or stains to the fluxes:

Cobalt oxide for blues
Copper oxide for turquoise and some greens
Iron and gold for reds
Platinum for grays
Manganese for purples
Iron for browny reds and some greens
Uranium and antimony for yellows
Tin for white
Iridium for black

These mixtures usually produce what is called a "transparent enamel." Opaque enamels are made by adding the same oxides to produce the color, and then by the addition of more tin oxides, antimony, and clay. It is possible to buy lead-free enamels, which are especially suitable for use by children, but the colors are generally not quite so varied as the standard enamels, and they do not mix well when laid with other enamels.

The temperature at which enamels fuse to the metal varies according to the color and whether it is transparent or opaque. The melting temperature depends on the proportions used in the composition of the enamel, and when you buy enamels you will see that each one is identified by a number and whether it is hard, medium, or soft. "Hard" means it has a high firing temperature; "medium" indicates an average firing temperature; and "soft" means that the enamel has a low firing temperature.

To begin with buy a range of simple colors, which should include a couple of blues (always an easy color), light red, green, yellow, white, and black. White and black enamels are always opaque, but the other colors can be transparent or opaque. You will also need some clear fluxes, which are often used as a base coat for transparent enamels and also as a final glaze. It is better to see an enamel color chart at your suppliers than to rely on a printed chart, but, whichever way you choose your color, remember that it will not always end up exactly as you imagine, because the color

depends on what metal you use and if a flux or a white base is used. The color you see of the enamel when you are buying it, in either lump or powdered form, does not always bear much resemblance to the fired color.

The identifying number written on all enamel containers is very important. It helps when you reorder a color, and is useful for keeping records.

Enamels are available in three different types – transparent, opaque, and opalescent. Transparent enamels allow light to pass through them to reflect the surface beneath. The brighter the surface of the metal before enameling, therefore, the brighter the enamel will appear. The density or tone of a color deepens with successive applications.

Opaque enamels completely cover a metal surface so no light is able to pass through it. After firing they have a shiny appearance.

Opalescent enamels give a slightly milky appearance, which allows some light to be reflected from the surface of the metal. They require very accurate firing to achieve their pearl-like appearance.

You can buy these three types of enamel in lump, frit form, or powdered.

Lump or frit enamels look like small lumps of colored or clear glass. Lump enamel is stored in airtight containers to prevent deterioration, and it should be absolutely dry before storage. I like to store the lumps in glass jars for quick identification, but the most important thing is that the jars have tight-fitting lids, which do not rust. If rust particles get among the enamel, it is very difficult to clean.

When it is stored correctly, lump enamel will keep for several years.

Selection of lump or frit enamels.

Before it can be used, however, you will need to grind the lumps with a mortar and pestle.

Powdered enamels have been commercially ground and packed. They do not usually need any further grinding, but they should be washed before use. They should be stored in a dry place and in an airtight container, but even so they deteriorate more readily than lump enamel.

I recommend that you use powdered enamels to start with – they are quicker to use, and you will soon become familiar with the feeling of them when you grind and wash them.

Enamels should look a little like fine salt before they are laid on the metal for firing. Both powdered and lump enamels contain some impurities, and they must be washed before use.

Palette with a selection of powdered enamels.

WASHING AND GRINDING LUMP ENAMELS

You will need
Mortar and pestle
Lump enamel
Tap water
Hammer
Distilled water

If you have to break up a large lump of enamel, break it as described here, remove the unneeded pieces from the water, and dry them very thoroughly before replacing them in the storage jar.

1 Place a piece of lump enamel in the mortar, and fill it about half-full with water.

2 Break up the lump by holding the pestle over it and hitting the top of the pestle with a hammer.

3 Hold the pestle firmly and use a circular and side-to-side motion to break down the lump. You will feel it beginning to become fine. Continue grinding until it is smooth, or until you can see that it has broken down smoothly.

4 Gently tap the side of the mortar with the handle of the pestle to help the enamel settle at the bottom.

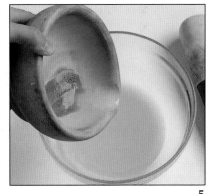

5 Taking care not to pour out the enamel, pour away the water, which will look cloudy. Half-fill the mortar with fresh water.

6 Hold the mortar in both hands and swirl the water around. It will still be cloudy. Tap the side with the pestle to help the enamel settle, and pour off the water.

7 Continue rinsing the enamels until you pour away really clear water.

8 Give the enamel a final rinse with distilled water. The enamel is now ready, either to be laid "wet" onto the metal or to be allowed to dry for sifting onto the metal.

Gallery

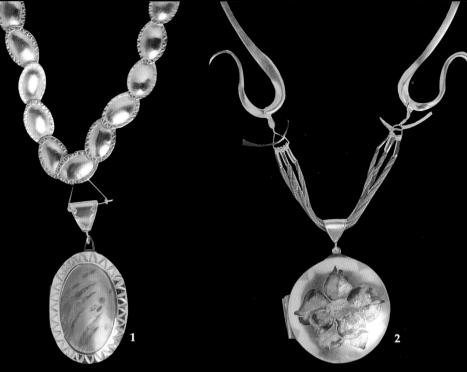

1. "Underwater" silver and basse-taille enameled locket and chain.
JINKS McGRATH.

2. Silver, 24-carat gold, and champlevé enamel locket hung on kumihimo silk braid and silver choker.
JINKS McGRATH.

3. Sheffield Park silver and cloisonné enameled locket, hung on cast silver and enamel chain.
JINKS McGRATH.

1

2

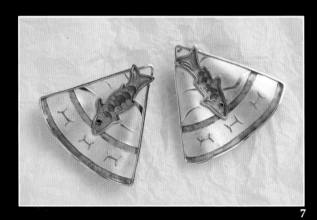

7

6

4. Repoussé silver cross with gold cloisonné enamel hung on handmade chain.
JINKS McGRATH.

5. Silver and enamel "ladybug" locket on silver chain.
JINKS McGRATH.

6. Champlevé and cloisonné silver earrings.
JINKS McGRATH.

7. Cloisonné and champlevé "fish" earrings.
JINKS McGRATH.

8. Hand-formed fine silver
beads, silver cloisonné
enameled/reverse pattern
"feathers." Silver and onyx
beads.
ALEXANDRA RAPHAEL.

9. Fine silver, hand-formed
shells, gold cloisonné wire
inside, and 18-carat gold wire
dangles.
ALEXANDRA RAPHAEL.

WASHING AND GRINDING POWDERED ENAMELS

You will need
Powdered enamel
Screw-top jar
Tap water
Distilled water

1 Powdered enamels can be washed by placing the desired quantity into a screw-top jar and covering well with water.

2 Stir or shake the bottle, and allow the enamel to settle on the bottom. Impurities will rise to the top, and the water will be cloudy.

4 Continue shaking until there are no more impurities and the water is clear. Give the enamel a final rinse with distilled water.

Alternatively, powdered enamels can be washed and ground using a mortar and pestle. Tip the required amount of powdered enamel into a mortar and cover with water, which should come about half-way up the mortar.

Use the pestle in the same way as for lump enamel – use a round and round, side-to-side motion – but this time you are not really grinding but washing the enamel thoroughly and releasing impurities. If you require very fine powdered enamel, continue grinding until it is the right grade.

Pour off the cloudy water and refill the mortar with water, swirling it around to rinse through the enamel.

Tap the side of the mortar with the pestle handle to settle the enamel and pour off the cloudy water.

Continue rinsing like this until the water is completely clear.

Give the enamel a final wash with distilled water. The enamel is now ready for use.

3 Pour away the water, and refill the jar with fresh water.

APPLYING DRY ENAMELS

You will need
Mesh stand
Firing fork
Sheet of clean paper
Gum
Sifter or sieve
Enameling tools

Drain away all the water after the final rinse with distilled water. Either leave the enamel in the mortar or place it in a clean, flat dish. Cover the mortar or dish with a clean piece of cloth or paper towel, and leave it to dry near to, or on top of, the warm kiln. Stir the enamel once or twice to help break down any lumpiness. You will know when it is completely dry because the grains will separate and the color will appear much lighter. It is then ready to be sifted onto your work.

Prepare the metal, then, without touching the clean surface with your fingers, place it on a stand.

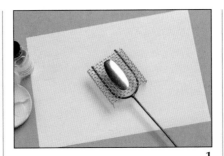

1 Use the fork to lift the stand into the middle of the clean piece of paper.

2 With a sable paintbrush, coat the surface of the metal with a fine layer of gum.

3 Put the sifter on the paper, and use either the little spatula or a teaspoon to take the dried enamel from the mortar or dish and put it into the sifter.

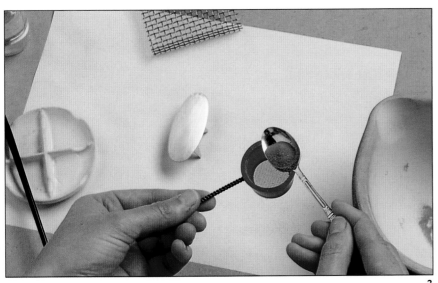

4 Hold the sifter over the metal, no higher than 3–4 inches, and carefully tap the side of the sifter with your finger, so that the enamel falls evenly over the surface of the metal. Try not to let the enamel build up in the center of your work – it is better to have it slightly higher at the edges than at the center.

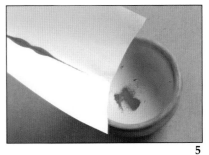

5 Remove the work on the stand from the paper, and carefully fold the paper into a funnel so that you can return the enamel to the mortar or dish.

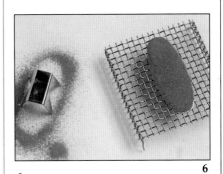

6 If you have enough enamel on your work after the first application, leave it to dry for a few minutes, and it is ready to fire. If it needs a little more, repeat the process, returning any unused enamel to the mortar for reuse.

1

2

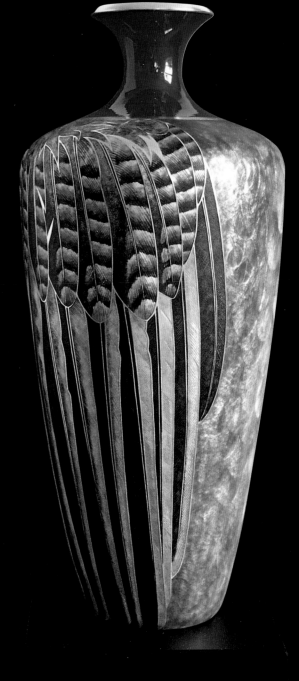

3

1. 18-carat gold box with green enamel and carving by Malcolm Long.
GERALD BENNEY.

2. 18-carat gold box set with rutilated quartz carving by Malcolm Long.
GERALD BENNEY.

3. Jay's wing vase – champlevé on silver.
JANE SHORT.

4

5

6

4. Silver, enamel, and fine gold brooch with chased silver detail.
SHEILA McDONALD.

5. Sterling silver brooch with transparent and opaque enamels with fine silver cloisonné wires. Top section is sandblasted slate with gold foil.
JESSICA TURRELL.

6. Sterling silver brooch with transparent and opaque enamels and silver cloisonné wires.
JESSICA TURRELL.

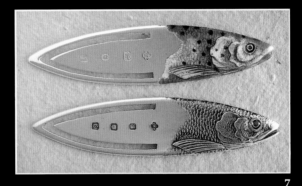

7

7. "Fish" bookmarks. Die-stamped silver form with underglaze enamel oxides and transparent colors on top.
JOAN MACKARELL.

8. Black and white enameled vases on sculptured copper foil.
MAUREEN CARSWELL.

8

USING A CONTAINER

You can buy jars with mesh tops for sifting dry enamel over your work, or you can make your own as described on page 14.

1 Place the dried, powdered enamel in the container, and screw down the lid firmly.

2 Lift the stand with the prepared metal into the center of a clean piece of paper.

3 Use a sable paintbrush to apply a fine coat of gum over the surface of the metal.

4

4 Hold the container no more than 3–4 inches over your work, and sprinkle the dry enamel evenly over it. As with the sifter, try not to let the enamel build up in the center of your piece. The first layer of enamel before firing should cover the metal well. If it is too thin, the enamel will "draw up" in the kiln, leaving little holes and bare metal. If the enamel is allowed to build up in the center, it will probably flake off after firing. If it is too thin at the edges, it will draw away from the edges. Save excess enamel in a dry jar with an airtight lid. Wash before reusing.

LAYING WET ENAMELS

You will need
Paper towel or mesh stand
Gum (optional)
Paintbrush
Enameling tools

When you have given the enamel a final rinse with distilled water, leave just enough water at the bottom of the mortar to cover the enamel. Swirl the mortar around, then tip it up and pour the enamel and water into a palette. Try to do this in a single movement, so that it all goes into the palette at the same time. Scrape any remaining enamel from the mortar into the palette, then wash the mortar thoroughly so it is clean before you grind more enamel.

1 Prepare your piece of metal, and place it carefully on a clean sheet of paper towel or on a mesh stand. Take care that when you handle the metal you always hold it by the edges. Try not to let your fingers touch the surface of the metal, because a trace of grease from your fingers can make the enamel draw away from that area as it is being fired.

2 If you are enameling a curved surface, you can apply a thin layer of gum with a sable paintbrush. However, gum is not really necessary when you are using wet enamels.

3

3 Put your little spatula or goose quill into the enamel in the palette, and drag it out of the water so that you have a tiny amount on the end of your tool. The solution should be wet, but there should be more enamel than water on your spatula.

4

4 Place the enamel on the metal and spread it out carefully. Continue to take little amounts of enamel and lay them on the surface until you have an even covering. The enamel should be as thin as possible while still covering the metal. Use your handmade spreader to help the enamel sit evenly.

5 Tap the side of your work with the handle of your spatula to encourage the enamel to flatten and spread.

6 If you have too much water on your work, which will make it difficult for the enamel to lie smoothly, hold the corner of a piece of absorbent paper just up to the side of the enamel – not on top of it – to draw off the excess water.

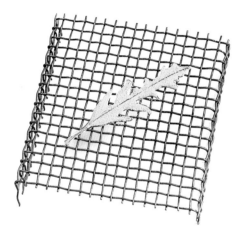

7 Continue laying on the enamel until you have covered the surface, then leave it to dry completely before firing.

You may find it difficult to lay wet enamel next to enamel that has dried, because the dry enamel takes away the water before you have a chance to lay the wet enamel. You can dampen your work with a small water spray, or by laying a little distilled water on the area you want to cover before you apply the enamel.

FIRING ENAMELS

Most enamels will fire at temperatures between 1,400°F and 1,650°F. The actual temperature at which each enamel vitrifies or becomes glossy depends on whether they are hard-, medium- or soft-firing enamels. The harder the enamel, the higher the firing temperature. The time taken will depend on the size and thickness of your work, and on the enamel used. A large piece could take 3 or 4 minutes to fire, while a small one could take less than a minute.

You can assess the correct firing temperature and timing for your enamel by testing the colors you wish to use on a sample of the metal you will be using. Enamelers fire a test piece to:

- Assess the firing temperature of chosen enamels.
- Test for color, over a metal and over a clear flux.
- Assess the suitability of an enamel for the piece and to check if the colors look good together.
- Assess how each enamel reacts to acid (used when removing oxides from metal) and to polish, because some enamels pick up dirty, greasy marks from polishing, which are impossible to remove.

For firing specific test samples see page 32.

FIRING A PIECE

When the enamel is completely dry, it is ready to be fired. It is very important at this stage to make sure that no particles of dust or dirt have somehow landed on your enamel. If you spot any dirt, moisten the end of a fine paintbrush and use the tip to lift it off, taking care not to disturb the dry enamel. While enamels are drying, try to make sure that they are kept in a place that is free from drafts and from obvious dust particles.

Enamel dries more quickly if it is held at the mouth of the kiln for a few seconds.

You can help a "wet" enamel to dry by holding it with a long firing fork on a wire mesh at the mouth of the kiln, or it can be left to dry naturally in a warm area near the kiln. You can remove any excess water with the edge of a paper towel placed just up to the enamel. If enamel is placed in a kiln before it is dry, the water boils quickly and evaporates, leaving little holes in your fired enamel.

5 STAGES OF FIRING

1. Enamel is dry, but has not adhered properly to the metal.
2. Enamel has started to fuse, but is uneven.
3. Enamel at the "orange peel" stage.
4. Enamel has fired, but is not yet smooth.
5. Enamel fired at the correct temperature and for the correct length of time to make it smooth.

FIRING A PIECE

1 Put on your protective glove, and open the kiln door.

2 Place the firing fork under the wire mesh on which you have placed your piece, and gently place the mesh in the kiln. Remove the fork.

3 Close the kiln door. Some kilns have a peephole in the door so that you can keep an eye on your work. If yours does not, open the door just enough for you to assess its progress. You can leave the door open during firing, but the kiln will lose a fair amount of heat and to compensate for this, you will need to set the kiln at a higher temperature.

4 As the enamel starts to melt, you will see that it turns darker. As it heats up it takes on a mottled appearance, and this is followed by a ripply shine. If it is left longer in the kiln the ripples will flatten out, but this flattening is usually left until the final firing. Remove the work from the kiln with the long firing fork.

5 Place your enameled work on a heat-resistant mat, which should be placed near to the kiln so that the enamel can cool down slowly.

Enamel fired to "orange peel" stage.

6 When it is cool, the enamel can be pickled to remove the oxides that will have formed on the metal in the kiln. Check your enamel is safe in the pickle by referring to your sample.

7 After pickling, rinse your work well with water. I sometimes rub pieces gently all over with the glass brush to make sure they are clean.

8 Apply the second coat to your piece, and refire.

9 If applicable, build up the enamel in several very thin layers, and fire between each layer.

10 On the last firing, allow the enamel to flow until the surface is smooth.

STONING DOWN

After firing, enamels may look uneven. Stoning down gives a good finish. For this process, the enamel must be slightly higher than the surrounding metal. Hold the piece in one hand under running water and rub the carborundum's flat side evenly across it. Continue, taking care not to grind away metal, until the enamel and metal are level. Wrap wet and dry paper round the stone, hold it under water, and rub across the work until you have a fine finish. Work through graded wet and dry from 240 to 600. Rinse thoroughly, and dry. Give the piece a flash firing to restore shine.

FLASH FIRING

This requires a slightly higher temperature than previous firings, so remove the work from the kiln as soon as the enamel is smooth and shiny. If an enamel needs several firings, you may not need to pickle between them. Any flaky firescale should be removed from copper before a new coat of enamel is applied, but silver may not oxidize after pickling.

FLUXES

There are hard, medium, and soft fluxes, each type compatible with different enamels. Before you begin a piece, check the compatibility of the fluxes and enamels you plan to use by testing them on scrap metal.

A hard or high-firing flux is usually used as a base coat under transparent enamels on copper. It is also a good counterenamel because it stands up well to firings.

A blue flux is often used to enhance transparent reds and pinks because it fires a silvery color compared with some fluxes, which look yellowish. A hard silver flux is a good base for cloisonné. Used carefully, it is compatible with medium; and soft-firing enamels.

Soft fluxes can be used as a protective layer over an enamel, and to fill up cloisons (cells formed by cloisonné wire) over colors that need one layer. They also prevent the color becoming too dark when several layers are built up. A flux can also be used all over an enamel to give depth.

The first firing with flux is usually at a slightly higher temperature than subsequent firings. If pickling is needed after the final firing, a flux can be used to cover the final layer of enamel. Some enamels react badly to pickling in acid. In this process, the flux protects the colored enamel from the acid.

COUNTERENAMEL

Counterenamel is used to relieve the tension that is created when enamel is fused to one side of a piece of metal. The enamel and the metal cool and contract at different rates, and this causes the metal base to curve. If more layers of enamel are applied to the metal without relieving this pressure, they will "ping off," or crack, when cooling.

When a layer of counterenamel is fired to the back of the metal, it evens out the tension, and this allows the metal to remain flat and the enamel to stay in place.

Counterenamel is sometimes difficult to work into the design of a piece. If this is the case, tensions can be overcome by altering the shape of the metal, to make a dome-

like shape, and then laying a layer of enamel onto the outside of the dome and omitting the counterenamel on the inside. If a piece comes out of the kiln slightly curved, it can be flattened, while still hot, by placing it quickly onto a warm, flat metal surface and placing another flat metal surface on top of it.

The need to apply counterenamel can be avoided by using a thicker piece of metal than usual. If, for example, you normally use a flat piece of copper or silver that is 0.5mm thick, you would need to counterenamel it to keep it flat by applying a layer of enamel 0.2–0.3mm thick on the front. However, if you used a flat piece of copper or silver about 1.2mm thick, you could lay on the same amount of enamel without needing to counterenamel.

METALS FOR ENAMELING

Copper

Copper is a soft, malleable metal that is well suited to enameling, and because it is comparatively cheap, it is an ideal metal for beginners to the craft. Copper can be shaped and bent easily, it can be cut with metal cutters or bull-nosed metal cutters, and it can be etched without being too extravagant for champlevé enameling.

Copper has a melting temperature of approximately 1,900°F, which is higher than the enamels, so there is less chance of spoiling a piece by melting it. Copper is very often used as a base for enameled jewelry, and, when enameling is completed, the separate pieces can be set into a silver or gold surround. It is also used extensively for larger pieces, such as bowls, panels, plates, and boxes.

When copper is heated an oxide forms on the surface, which, if left untreated, will spoil most enameled work, although it is sometimes used to advantage. Normally, the oxide,

which is black and flaky, must be removed before each firing, by either immersing the piece in a "pickle," or by scrubbing the surface with steel wool, followed by wet and dry papers.

Annealing The process of annealing involves heating metal to a certain temperature so that, when it is cooled, the metal will be soft enough to work with. The process also cleans any dirt or grease from the metal. You can anneal copper by placing it in the kiln, or with a torch.

If you use a kiln, preheat it to about 1,300°F. Put the copper on a mesh tray, and place it in the kiln. Leave it for a minute or two, until it looks red, remove, and cool immediately in water. Clean the surface of the copper by immersing it in your pickle.

If you use a blowtorch, place the copper on a soldering block in a soldering tray. Work against a dark area that allows you to see the color of the flame of your gas torch. The heat of the flame is in the orangey-blue part, just behind the tip of the

flame, and you should play that soft blue flame all over the piece of copper. As you do this the surface of the metal will start to change color. Continue until it becomes red. Hold the metal in the flame until the whole piece looks red in appearance for a few seconds, then put out the flame and cool the copper immediately by dipping it in water. Use brass tongs or tweezers to lift the copper from the heating area to the water. Finally, clean the surface of the copper by immersing it in a pickle.

Pickling To clean the surface of the copper after annealing, the piece can be immersed in a solution of 10 parts water to one part sulfuric acid; or 2½ cups vinegar with 4 tablespoons salt.

Leave the copper submerged until the surface is completely clear. The vinegar solution is the safest and easiest if you do not have a separate work room, but it takes a little longer to work well than the acid solution.

When it is clean, lift the piece out of the pickle with brass tongs, rinse it well under cold water, and prepare the surface for enameling.

Preparing copper for enameling It is easier to prepare all surfaces for enameling by keeping them wet. Take a piece of wet and dry paper – I use grade 240 to start with, and then, if necessary, work down further with grades 400 and 600 – and rub it over both sides of the

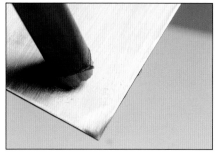

Use a glass brush to give the copper a final clean.

copper. Then use a glass brush all over the surface until the metal shines, and the water is able to stay smoothly on it. If the water pulls away or draws up into globules, the surface is not yet clean enough. Remember that the enamel will behave much like the water and will pull away from the surface during firing if the surface is not sufficiently clean.

After thorough cleaning, place the copper carefully onto a clean tissue, paper towel or stand, and do not touch the cleaned surface.

A good way of giving a final clean to the surface is to lick it, because saliva is a good neutralizer.

Finishing a piece After its final firing the surface of the enamel will look rich and shiny, and no further work is required, unless you want a matte finish. The metal will, however, usually need cleaning and polishing after enameling. The black metal oxides that form on the metal while it is in the kiln can be removed by pickling the piece in one of the solutions detailed under "pickling" on this page. The oxides will disappear fairly quickly in the sulfuric solution, but will take a little longer in the vinegar.

Some enamels react badly when they are immersed in acid, and you should always test for this *before* putting a finished piece in the pickle, or you may find out too late that the enamel did not like it.

Sulfuric acid can be ordered through a friendly pharmacist, and it is possible to get a "safety pickle" from any jewelry tool suppliers. Remember that **acids are dangerous,** and they should always be handled with great care. Use rubber gloves, and always add acid to water, not the other way around.

After pickling and rinsing, the metal can either be polished by hand or by machine. If you are using a machine, use a felt lap with a slurry of 240 mesh pumice powder

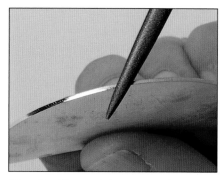

Use a fine file to polish the edge of a piece of silver.

mixed with water or glycerin. The electric motor should run between 900 and 1,200 rpm. Follow with a final polish with rouge (polish agent) on a lambswool mop. Again, test enamels before polishing, as some immediately take up the grease of the polish and become impossible to clean.

Hand polishing can be done in several ways. Metal edges can be cleaned successfully with a fine file, followed by rubbing with a polished burnisher. Liquid metal polishers can be used with a chamois leather cloth or a duster. You can polish flat areas with a buffing stick, which is nothing more than a piece of flat wood, about 1 × ½ × 12 inches, with a piece of felt or leather stuck to half of it. Use the pumice slurry on the felt stick, and the rouge polish dissolved with a little lighter fuel on the leather stick.

Enamels can also look very attractive with a matte finish. After stoning down, work carefully through each grade of wet and dry papers, finishing with 600 or 1,200. The piece does not then need refiring, and you can polish the metal separately if you want. You can achieve a similar effect by holding the finished enamel, after the final firing and polishing to the metal, in matting salts. These bite into the surface of the enamel and give it a slightly frosted look. Rub a little silicone wax onto the enamel to protect against dirt.

Silver

There are four types of silver to choose from for enameling: fine silver is 999.9 parts per 1,000 pure silver; Britannia silver is 958 parts per 1,000 silver; sterling silver is 925 parts per 1,000 silver (the rest is copper); and sterling silver/ enameling quality, which is the same proportions as standard sterling silver, but the copper and silver are combined to a very high purity.

Fine silver has a beautiful color and is ideal to enamel on. Unfortunately, it is very soft, so it is generally impracticable to use it for jewelry. However, it can be enameled as a separate item and then set into a sterling silver surround. The purity of the metal means that no firestaining occurs when it is heated.

In theory, Britannia silver should be better than standard silver to enamel because it has a higher silver content. However, it too suffers from being rather soft, and does not have the advantage of the very pure color of fine silver. It usually has to be ordered specially.

Sterling silver would be perfect metal to enamel if it did not suffer from firestain. Unfortunately, after heating, the copper content in sterling silver stays on the surface in darkish grey patches, which are called firestain. Enameling over firestain causes transparent enamel to become dull and cloudy, and although opaque enamels are not affected, one does not usually want to cover the whole of a silver surface with opaque enamels. Apart from the firestain, annealed sterling silver is soft enough to work with, and hard enough to be practicable for jewelry purposes.

Sterling silver/enameling quality is basically the same as sterling silver, with the same firestain problems, but the copper and silver are of very high quality, which eliminates impurities.

Annealing Preheat the kiln to about 1,110°F, and place the silver on a mesh stand in the kiln. When the silver is a dull pink all over, take it from the kiln and cool it quickly in water. Clean the surface by pickling.

If you are using a torch, work in a darkened area and play the end of a soft blue flame over the silver until it turns a dull, pinkish color. Continue to heat the silver until it is this color all over for a few moments, then put out the flame before cooling the silver quickly in water. Clean the surface by pickling.

Torch flame with "blown" air – correct for annealing.

Pickling Cleaning the surface of silver is much the same as cleaning copper, the more usual solution being 10 parts water to one part sulfuric acid. A solution of 2½ cups vinegar and 4 tablespoons salt does work, but it takes time! Place the silver in the solution with brass tongs, and let it sit there until the surface is clean. Remove with the tongs, and rinse well. The pickle will work more quickly if the solution is warm. Keep the acid or vinegar solution in a glass container, place the container in a pan or pyrex dish containing water, and heat the water. You should keep a lid over the pickle.

Silver, half immersed in "pickle" after heating.

29

DEALING WITH FIRESTAIN

The easiest way to deal with firestain is to bring the pure silver in the sterling silver to the surface. If you anneal, then cool and pickle the silver, you will find that the surface gradually becomes whiter, which is the fine silver coming to the surface. You will need to repeat the annealing and pickling process four or five times to get a really good coat of fine silver.

Alternatively, you can paint the surface of silver with a proprietary mix such as Argotec, a soldering flux mixed to a paste with water or methylated spirits and painted on the silver before any heat treatment. After annealing, clean off any residue, and reapply when the silver needs further heating. Take care not to paint soldering flux on any soldered joints. It is essentially a flux and would cause the solder to run.

It can be used while you are enameling, but take care that none of the paste gets on the enamel.

If you fail to prevent firestain occurring, you can get rid of it in a solution of three parts water to one part nitric acid. Use brass or plastic tweezers to immerse the silver in the solution for a few seconds. Look out for a black area appearing on the silver. As soon as this appears, lift it out of the acid and hold it under the running water, removing the black area by rubbing it with pumice powder

Argotec is painted onto a ring to prevent firestain.

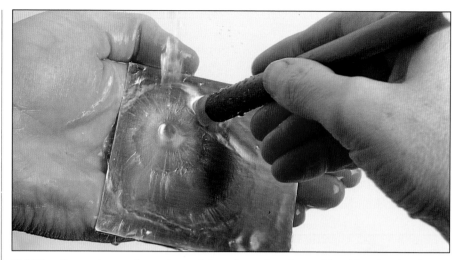

Hold the silver under running water, and clean the surface thoroughly with a glass brush.

mixed to a creamy paste with water. Use a soft toothbrush to apply the paste. Continue to do this until the black patch has disappeared. Be extremely careful if you use this method when you have any soldered joins, because the solution is an etching solution and will quickly attack soldered joins and weaken them.

Alternatively, you can buy a stick of soft abrasive compound (called water of Ayr stone). Rub this stick, which must constantly be kept wet, over the area affected by the firestain and work it into a paste and the stain will fade. Check your progress by washing away the paste. Continue until the stain disappears.

Preparing silver for enameling The shinier the surface of silver under transparent enamel, the more reflective it will be and the better the enamels will appear. Thoroughly rinse the silver with water, and clean it with the glass brush until the water stays all over the surface without drawing up into globules. Any surface work, such as texturing, engraving or etching, should be completed before the final cleaning. Place the silver on a paper towel or stand, and lick it (see Preparing Copper) before you lay on the enamels.

HAVING FUN

There are other ways in which you can enamel on copper. These include using a swirling or scrolling tool, firing in millefiori, enamel threads or lump enamels, stenciling, and scraffito. All these techniques, with the exception of some scraffito, involve using a hard flux or hard white base coat, and the piece should always be counterenameled. Sift a layer of transparent colored or flux enamel onto the piece before placing any decorative media on it. Millefiori are little, roundish lumps with patterns going right through them, making them somewhat resemble rock candy! Enamel threads are rods of enamel, which can be laid in lines or broken into little pieces, and lump enamel, in either transparent or opaque form, will fuse and stay slightly proud unless it is allowed to fire a bit longer and at a slightly higher temperature, whereupon it fuses flat.

Millefiori, Threads, and Lump Enamel After sifting a transparent color over the previously fired base coat, place the decoration into the sifted enamel, tucking it into the enamel so that it is held while it is being fired. If you are working with a curved surface, dip the decoration in a little gum before placing it into

Dip the lump of enamel into gum before placing it into dry sifted enamel with a pair of fine tweezers.

the enamel. Millefiori, in particular, needs quite a high temperature before it will fuse. The project to decorate a nut bowl involves placing lump enamel. If you want the millefiori to lie flat, allow it to fuse, and then leave it in the kiln slightly longer than usual to allow the surfaces to spread.

Stenciling You can have a lot of fun with stencils and enamels. Use any bits and pieces you have to hand as a stencil – your wire mesh stand, leaves, and petals, for example. Try sifting a transparent enamel onto your base coat through a stencil, and then move it a little so that the next sprinkling of enamel falls over a corner of the first area. This will produce a variety of attractive color tones, while building up an interesting pattern.

Swirling or scrolling Sift a layer of colored transparent enamel over a fluxed or base white coat. Take little pinches or spoonfuls of another color enamel, and sprinkle it in the areas you want your swirling pattern to be. Place in the kiln and fire. At the point you would normally remove it from the kiln, open the kiln door and prop it open. Hold the piece steady with one swirling tool, and place another tool into the molten enamel and trail a swirly pattern into the two

colors. You will need to be very quick doing this, otherwise you will lose heat in the kiln through the open door. When you have finished swirling, close the door of the kiln to allow the swirled colors to fuse flat. Drop your tool into cold water to remove any enamel that has stuck to it. Remove the piece from the kiln and cool. If you want to swirl any lumps, threads or millefiori, the kiln will need to be hotter. Wait until they have fused into the transparent enamel layer, swirl them with your tool, leave them to fire a little longer with the door shut, and then remove them.

If the tool gets stuck in the enamel, it is because the kiln has cooled down too much while the door was open, and the enamel has become less molten and is sticky. Try shutting the kiln door as far as possible until it heats up enough to release the tool. You must remember to swirl as quickly as you possibly can.

Scraffito Scraffito was originally a Japanese technique of tracing a design onto a piece through a color so that the traced lines appear as another color. There are various ways of achieving this.

You can simply sift an enamel – probably an opaque one – onto the copper, which you have counterenameled. Use a pointer to draw a pattern through the sifted dry enamel, removing any obvious bits of enamel left in the lines with a

moistened paintbrush, as is done in the project to make a set of coasters. Fire the piece. The copper will then need cleaning because firescale will have occurred in the kiln. The shiny copper will show through the enamel as a bright line.

Alternatively, you can enamel a base coat onto the copper and counterenamel it. Then sift another color over the fired base coat. Draw the pattern through the top sifted layer of enamel, and fire the piece. The lines of the pattern will show through the top color in the color of the base coat. To make the surface even, you can then lay in the color of the line, by either using wet enamel or by gumming the line and sifting on the color. Remove any stray enamel with a moistened paintbrush as in the project to make a pendant with copper wire decoration, and then fire until it fuses flat.

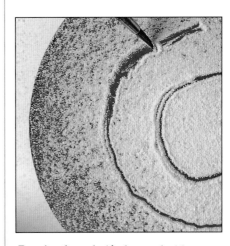

Drawing through sifted enamel with a paintbrush to produce a "scraffito" effect.

Silver brooch with cloisonné wires and "scraffito" drawn pattern.
JINKS McGRATH.

Doing Tests

Carrying out tests is one of the most important aspects of successful enameling, for the quality of a finished piece can largely depend on what you have learned from your tests. You are aiming to discover the stage at which each enamel fires – that is, whether they are hard, medium, or soft; what difference a flux makes underneath a transparent color; to test whether the enamel is sensitive to immersion in acid; to see if polishing affects it; and to put different colors side by side to assess their compatibility for a piece.

If you are methodical about your test pieces, they will be an invaluable reference before you start any new work.

Simple sample

1. Clean a piece of copper or silver about 1 × 1½ inches, and then use a sharp steel point to scribe three parallel lines across one side. Drill a hole in the top.
2. Counterenamel the back with either hard white or flux.
3. Clean the front by pickling it in sulfuric acid or a solution of vinegar and salt, rubbing with wet and dry papers, and finally with the glass brush to leave a perfectly clean surface. Dry the work.
4. Leave the first section free from enamel. Place the hard white enamel in the second section. Place the hard or medium flux in the third section.
5. Fire the piece, and allow it to cool down.
6. Clean the exposed metal section.
7. Lay the colored enamel on top of the three sections, and fire the sample.
8. Scratch the number and make of the enamel on top of the sample, and make a note of firing times, temperatures, acid, polishers used, and so on.

Comprehensive sample

1. Clean a piece of silver or copper about 1 × 1½ inches and then use a sharp steel point to scribe five parallel lines across the front.
2. Clean the metal, and counterenamel the back with a hard or medium flux.
3. Clean the front of the metal by pickling, rubbing with wet and dry papers and shining with a glass brush.
4. Leave the first silver or copper section bare.
5. Put a hard white on the second section. Put a medium or a

Simple sample.

hard flux on the next two sections. Put a blue flux on the fifth section.
6. Fire the sample at a fairly hot temperature.
7. When it is cool, clean the bare metal section, then cover the first, second and third sections with the color, cover the fourth section with a strip of silver foil, and cover the fifth section with a smaller piece of gold foil.
8. Fire the sample, and allow it to cool down.
9. Put the color onto the fourth and fifth sections, fire, and cool. Add further color on all sections if you wish.
10. Scratch the number and make of the enamel on the top of your sample, and then make a note of firing times, temperatures, acid and polishes used and so on.

Comprehensive sample.

Blue Star Pendant

For our first project I have chosen a copper star shape to wear as a pendant. I used a powdered mazarine blue enamel, which covers copper well, and a medium powdered white enamel. The pendant hangs on a leather thong. As with all the projects in this book, thoroughly clean your work area before you begin, and make sure you have everything to hand.

You will need

Small hand drill
Copper star shape
Stand
Paintbrush
Gum
Teaspoon or small spatula
Powdered mazarine blue enamel –
    washed, dried, and ready for use
Sieve or sifter
Petri dishes for dry enamels
Pointer
Stand
Firing fork
Tile or plate
Powdered white enamel (medium) –
    washed, dried, and ready for use
Wet and dry paper, grades 240 and 400
Flat needle file
Jump ring
Leather thong

Template

1 Drill a hole about 1mm in diameter in the top point of the star with your hand drill.

2 Use a drill bit 2–3mm in diameter to countersink the hole in the star. Prepare the copper for enameling and rest it on the stand, taking care that your fingers do not touch the surface.

3 Cover the surface of the star with the gum, and use a teaspoon or small spatula to put some of the blue enamel into the sieve. Hold the sieve over the dish while you do this to catch the enamel as it falls through.

4 Hold the sieve about 3 inches above the star and tap it gently, making sure you cover the surface evenly. Put a little extra at the edges, because this is where the enamel might draw away if too little is applied. Put your pointer through the hole in the star to clear away any enamel. Carefully transfer the star to the mesh stand, and pick it up with the long firing fork.

5 Place the star in the kiln for firing. I had the kiln at about 1,560°F, and it took 50–60 seconds to fire. For your first few firing attempts, you could keep the door of the kiln open and watch the enamel as it fuses. Although it will take a little longer, the more you see how enamels behave in the kiln, the better.

6 Remove the star from the kiln with your firing fork, and leave it to cool on the tile in front of the kiln. When it is cool, place the star back on the stand, and paint a little gum onto the tips of the star.

7 Sift white enamel over the ends you have glued. Do not worry if a little falls in places you have not glued, but if there is too much, blow it gently away.

8 Refire the star, remove it from the kiln, and allow to cool. Clean the reverse of the copper with the wet and dry papers, and clean the edges of the star with a fine flat needle file.

9 Fix a jump ring through the hole, and close it up. Thread the leather thong through, and knot it at the back.

Earrings on Copper

This project involves using a base coat for the transparent color, called a flux, and we apply a counterenamel, which is described on page 27. Counterenameling is especially useful for dangly earrings, which can be seen from both sides.

You will need

Copper shapes for earrings
Stands
Paintbrush
Gum
Sieve or sifter
Powdered copper flux (hard) – washed,
    dried, and ready for use
Mesh tray
Firing fork
Wet and dry papers
Powdered, cherry red enamel – washed,
    dried, and ready for use
Flat needle file
Jump rings
Earring hooks

Template

1

1 Prepare the copper shapes for enameling, place them on the stand, and paint the top surfaces with a smooth coat of gum.

2

2 Sift the flux onto the surfaces of both pieces, applying as even a coat as possible.

3 Place the shapes carefully onto the mesh tray for firing.

4

4 Place in a kiln, preheated to 1,560–1,650°F, and fire until the enamel looks like orange peel (see page 26).

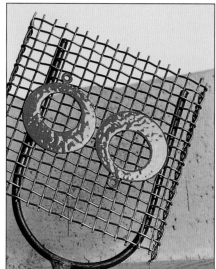

5

5 Remove from the kiln, and allow to cool down.

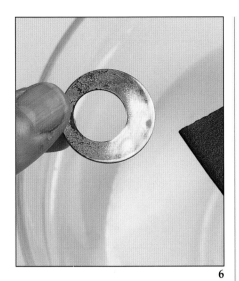

6

6 Clean the black oxides off the reverse with water and wet and dry papers.

7

7 Place the earrings, copper side up, on the stand, and paint the surfaces with gum. Sift a coat of flux onto each one. This is the counterenamel coat. Turn the supports the other way up, and place an earring between them so that only the outside rim of the earring touches it. Fire each earring individually until the enamel is smooth. Allow to cool, and then paint some gum onto the fronts of the earrings.

8

8 Sift the cherry red enamel onto them, then fire each one at approximately 1,500°F, or at a slightly lower temperature than you used for the flux.

9 Allow to cool, then clean up the edges of the copper with the wet and dry papers, taking care not to scratch your enamel. If any small pieces of enamel have fused to the edges, carefully remove them with a flat file and finish off with wet and dry papers.

10 Use jump rings to attach the hooks to the finished earrings.

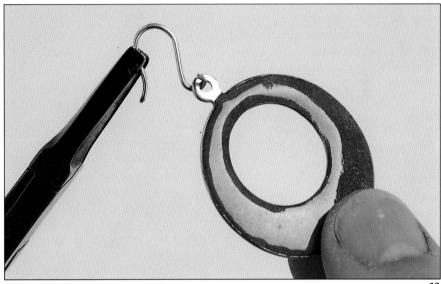

10

Earrings with Painted Decoration

These earrings are counterenameled, and enamel paint is used for the decoration. I used a simple green, but you could use several different colors, which can all be fired in a single firing. I have used pin and butterfly findings, but you could, of course, use a clip fastening if you prefer.

You will need

Copper sheet, approx. 0.6mm thick and 4 × 4 inches
Pickle solution
Wet and dry papers
Packet of copper ovals
Tracing paper
Pencil
Scissors
Spray glue or adhesive stick
Metal cutters or piercing saw
Sandbag or folded towel
Small flat file
Hand drill
Gum
Paintbrush
Powdered white enamel (hard) – washed, dried, and ready for use
Mesh stand
Firing fork
Glass brush
Powdered white enamel (medium) – washed, dried, and ready for use
Carborundum stone
Tube of enamel paint
Strip of glass
Jump rings
Contact adhesive
Pin and butterfly earring findings

Template

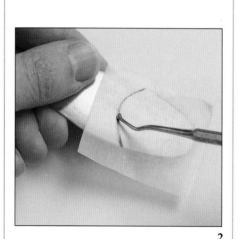

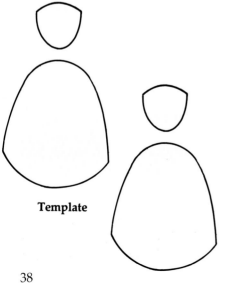

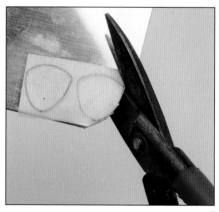

1   Anneal the copper sheet in the kiln at about 1,475°F for a minute or two to burn away any grease and soften the sheet to make cutting easier. Remove the copper and pickle it, then rinse the copper and clean it with wet and dry papers. Clean the copper ovals in the same way.

2   Trace the templates, and fix them onto the ovals for the lower half of each earring. Rub the pencil line so that it comes through onto the copper. Fix the templates for the upper halves onto the copper sheet.

3   Cut the ovals across the line with the metal cutters or piercing saw.

4   Draw the line for the other half of each earring on the remaining end of the copper oval, and cut out the shape with the metal cutters.

5   Cut out the two upper halves of the earrings from the copper sheet.

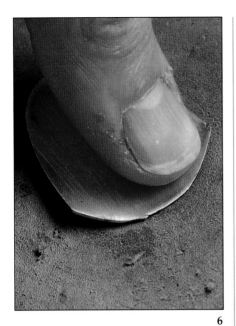

6 Shape the earrings by placing each of the four cut-out pieces onto the sandbag or folded towel and pressing down firmly in the middle with your thumb until you achieve the shape you want.

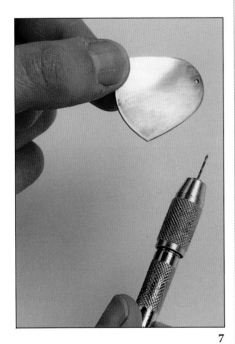

7 Neaten up the edges of the copper with a small, flat file, and drill a hole in the top of the two lower pieces and the bottom of the other two pieces.

8 Clean and prepare all the pieces to be enameled, then paint the underside of each piece with gum. Sift the hard white enamel onto the surface. Place all the pieces on the mesh stand, and fire in the kiln until the enamel is at the "orange peel" stage. Remove them from the kiln, and allow to cool.

9 Clean the copper on the front of the shapes with wet and dry papers, followed by the glass brush. Dry and paint with gum before sifting a layer of hard white enamel over each one. Support each piece between the legs of the stand so that the enameled backs are suspended. The enamel can be refired without leaving marks from the support. You will have to fire each piece separately, unless they will all fit on one support. (I usually place my supports on the mesh stand and lift the whole lot into the kiln with a long firing fork, because it is easier to hold steady.)

10 Remove the shapes from the kiln, and allow to cool. Paint a coat of gum onto the fronts, and sift on a layer of medium white enamel and refire. Remove from the kiln and cool.

11 Gently rub down the surface of the enamel on the front of the earrings with a carborundum stone. When it is flat, finish off by rubbing away the carborundum stone marks with wet and dry papers. This should all be done under running water.

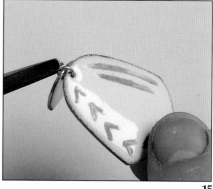

15 Clean up all the edges with a small, flat file, taking care not to scratch the enamel, then join the top and bottom pieces together with a jump ring.

12 Use a soft pencil to draw the pattern onto the enamel, which is easier than painting directly with enamels. If you make a mistake, wash off the line because an eraser can leave a smudgy mark. Dry the piece before redrawing. Refire the pieces so that they are shiny again. The pencil lines will be retained.

14 Place the pieces on the support stands, and allow the paint to dry. Hold the pieces at the mouth of the kiln, and if you see any evaporation you will know that the paint is not yet ready for firing. When the paint is perfectly dry, fire at the same temperature you used to fire the medium white enamel. It will probably take less than a minute, but make sure the white enamel has glossed over again. Remove and cool.

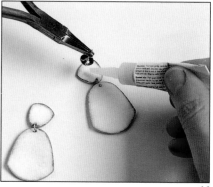

16 Use a contact adhesive to fix the posts to the back of the earrings.

13 Squeeze a little enamel from the tube onto the glass, and use a fine paintbrush to take a little of the color to paint between the pencil lines. Try not to put it on too thickly or you will get a matte finish.

Pendant with Copper Wire Decoration

This project uses copper wire as cloisonné to separate two colors and as a decoration. As with all the projects in this book, the colors I have used are only suggestions. If you should wish to use different colors, however, I suggest that you do a simple test as described on page 32 to assess their suitability and the necessary firing temperatures. I have used transparent colors, which allow the copper wires to fuse better than opaque enamels do. Opaque enamels usually take longer to fire completely, which allows a layer of oxides to build up on the copper wire. When the piece is cooled, this layer flakes off, which results in a lack of adhesion that may cause the wire to come out.

You will need

Copper sheet, approx. 0.6mm thick and 4 × 4 inches
Pickle solution
Wet and dry papers
Packet of copper ovals
Tracing paper
Pencil
Scissors
Spray glue or adhesive stick
Metal cutters or piercing saw
Small, flat file
Sandbag or folded towel
Wooden dome
Hammer
Hand drill
Wet and dry papers
Glass brush
Gum
Paintbrush
Copper flux (hard) – washed, dried, and ready for use
Sieve or sifter
Mesh stand
Firing fork
Round-nosed pliers
12 inches 0.8mm copper wire
Supports
Transparent "jaguar" green enamel – washed, dried, and ready for use
Transparent bronze enamel – washed, dried, and ready for use
Jump rings
Leather thong

Template

1

1 Prepare the copper for enameling, and trace the circle from the template for the top half of the pendant. Fix the tracing onto the copper sheet. Leave at least ½ inch of tracing paper outside the circle line when you attach it to the copper because this makes it easier to follow the line when you cut it out. Use your metal cutters to cut out the circle from the copper sheet.

2 File the edges of the circle to make them smooth.

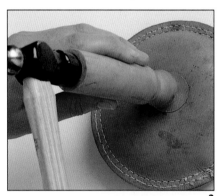

2

3

3 Put the circle onto the sandbag. Place the wooden dome in the center of the circle, and hit the top of the dome with a hammer. Move the dome around the circle to give a good, even shape. If you do not have a wooden dome, press the copper with your thumb. If it is annealed, it will be soft enough to shape.

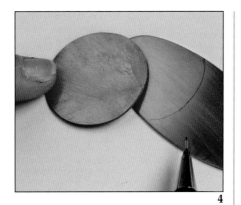

4 From the template, mark the lower half of the pendant on an oval shape. Place the domed circle up to the marked length, and draw the shape onto the copper oval beneath. Cut along the line with metal cutters.

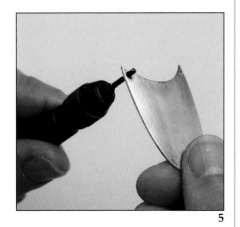

5 Drill holes in both the top and bottom pieces to accommodate the jump rings that will hold the pieces together.

6 Prepare the copper for enameling by cleaning it with wet and dry papers and a glass brush.

7 Paint the reverse side of both pieces with gum, and sift on the hard copper flux. Place the pieces on the mesh, and fire at about 1,560°F or higher until the "orange peel" stage.

8 Allow to cool, then clean up the fronts ready for the coat of flux. Support the two pieces on the stands so that the enameled backs are suspended.

9 Paint the fronts with gum, sift on the flux, and fire until almost smooth.

10 Use round-nosed pliers to bend the copper wires into the same shape as the pattern. Keep checking that the shape is correct by laying the wires along the lines of the template. Finish the wire for the top piece by bending up a double loop. You will be able to bend the wires between your finger and thumb to fit the curve of the two pieces.

11 Lay the wires on the pendant, using some gum to hold them in place, then place the two pieces on the supports. Place them in the kiln to fire until the wires just start to sink into the flux. Remove and cool. If any of the wires have oxidized, hold the piece under running water and rub them with a glass brush. Carefully push down any protruding wires – they will adhere in the next firing.

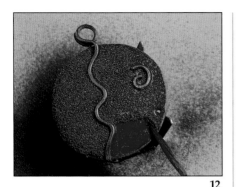

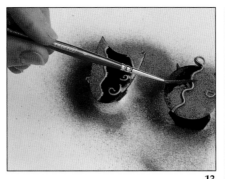

12 Paint the half of each piece that is to be green with gum, and sift the green enamel on that half. If any green falls onto the other half, just moisten a paintbrush and gently brush away the surplus enamel from the wires and the surface.

13 Fire the green enamel, remove from the kiln, and cool. Paint the areas to be enameled bronze with gum, and sift the bronze onto them, brushing away any stray enamel as before. Fire the piece until the enamel has a good, smooth finish.

14 Attach the two pieces with the jump rings, and place another jump ring through the looped copper wire at the top. Thread the leather thong through the jump ring, and knot it at the back.

Nut Dish

This little dish is easier to enamel than you might imagine. You can buy copper bowls at your enameling suppliers, but make sure that you choose a size that will fit into your kiln without touching any of the edges. The pattern is achieved by drawing the outline for each color separately in pencil, painting gum between the pencil lines, and then sifting the colored enamel onto the surface, where it is held by the gum. The transparent enamel is fired over white, which can be used instead of a clear flux as a base for transparent enamels. Put a ceramic kiln tile on the floor of your kiln for this project. When you work on a piece as big as this, it is difficult not to spill some enamel as it goes into the kiln, and you should try to keep the floor of your kiln as clean as possible.

You will need

Copper bowl
Gum
Paintbrush
Powdered white enamel (hard) – washed, dried, and ready for use
Sieve or sifter
Petri dishes
Kiln tile
Mesh stand
Firing fork
Transparent "jaguar" green enamel – washed, dried, and ready for use
Powdered white enamel (medium) – washed, dried, and ready for use
Small lumps of transparent gold enamel – washed, dried, and ready for use
Tweezers
Pointer
Diamond abrasive (medium)
Wet and dry papers
Pencil
Transparent pale pink enamel – washed, dried, and ready for use
Flat file

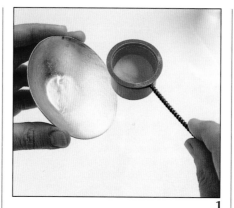

1 Clean and prepare the copper dish for enameling. Apply a good coat of gum to the inside of the bowl, then hold it in your hand and sieve the hard white evenly over the inside. You will find it easier if you tip the bowl towards you so that you sift one area at a time, finishing with the bottom. Empty excess enamel into a petri dish.

2 Place the bowl on the wire mesh stand, lift it up with the firing fork, and place it in the kiln. As it is bigger than work you have done in previous projects, you will find it takes longer to fire. Keep an eye on it, and remove it from the kiln as soon as the enamel starts to smooth out. Remove and allow to cool. Remove the black oxides from the underside of the bowl, and dry it thoroughly.

3 Place the bowl upside down on the wire mesh, cover the side surfaces with gum, leaving the center clear, and sift on the green enamel, making sure you apply a good covering all over. Brush away any excess enamel that falls on the center area.

4 Carefully place the bowl on the mesh stand, and put it in the kiln. Fire until the "orange peel" stage, then remove from the kiln and cool. Clean off the oxides from the bottom. Turn the bowl the right way up, and coat the inside with gum. Apply a second layer of white to the bowl, but use a medium white, sifting it in the same way as you did the hard white. Replace the bowl on the mesh stand.

5 Take 10 small lumps of transparent gold enamel, choosing pieces that are more or less the same size, and dip each one into some gum. Use your tweezers to push them into position around the top edge and the base of the bowl.

6 If necessary, use your pointer or other sharp tool to adjust the position, but try not to disturb the sifted white too much. Place the bowl in the kiln, and fire until the white enamel is smooth. Remove and cool. If the inside of the bowl looks rather lumpy now, rub it down with diamond abrasive under cold running water. Always use the wet and dry papers after the abrasive, because it leaves large scratches in the enamel, which do not always refire properly. Refire before going on to the next step.

7 Use a pencil to mark the areas that you want to enamel green. Paint inside one area with gum and sift on some green, carefully shaking or blowing away any excess.

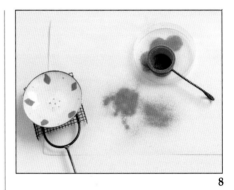

8 Continue gluing and sifting until all the green areas on the bowl have been covered. Fire until the green is smooth, remove from the kiln, and cool completely.

9 Use a pencil to mark the areas you want to enamel pink. Gum inside each section and sift on the enamel, removing any excess pink with a moistened paintbrush.

10 Place the bowl on the mesh stand, lift it into the kiln, and fire until it is smooth. Remove from the kiln, and allow to cool near the kiln.

11 Clean up underneath the bowl with wet and dry papers, and then carefully rub or file the top edge clean.

Coasters

We are going to use the scraffito technique for this set of coasters. You can buy copper circles ready to use, or you may prefer to cut them out yourself. If you are cutting out your own circles from a sheet of copper, you will need to anneal it before cutting and cleaning. File the edges of the circles smooth. The coasters are counterenameled, which is essential to keep them flat. I have described how to make just one – simply change the color each time to complete the set.

You will need

Copper circles, 2½–3 inches in diameter
Flat file
Wet and dry papers
Glass brush
Large stand
Gum
Paintbrush
Powdered opaque enamels in six
    different colors – washed, dried, and
    ready for use
Sieve or sifter
Firing fork
Pointer
Carborundum stone

1   Prepare the circles for enameling with the wet and dry papers and a glass brush.

1

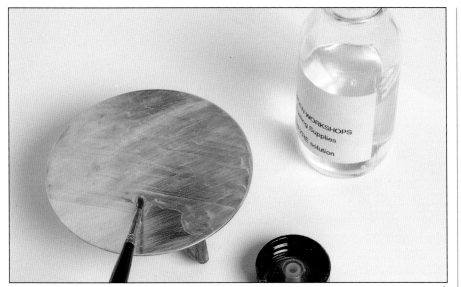

2

3

2   Place a circle on the stand, and cover it with a good coating of gum.

3   Apply the counterenamel first. I prepared a black enamel, which I used on all the coasters, but you could use the same color for the counterenamel as for the front. Sift on a good layer of enamel, making sure no copper is left exposed at the edges.

4

4 Place the coaster in the kiln, taking care not to spill too much enamel as you set it down in the kiln. Allow the enamel to fuse until the "orange peel" stage, then remove from the kiln and cool.

5

5 Clean up the front of the coaster with wet and dry papers, followed by a glass brush.

6

6 Paint a coat of gum onto the front, and sift on a good layer of the opaque enamel in the color of your choice. Place in the kiln, and fire until smooth. Remove, and allow to cool.

7

8

7 Sift on a fine coat of yellow enamel, putting more in the center and leaving the edges rather feathery.

8 Use your pointer to draw a spiral, starting from the middle outwards. Don't worry if it is a little wobbly – it is not designed to be precise.

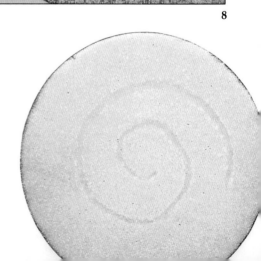

10 Place the coaster in the kiln, and fire until smooth. Remove, and allow to cool near the kiln.

9 If you want a wider line or need to clear away any enamel from the line, draw a moistened paintbrush through the line. Lift off any peaks that form.

11 Clean carefully around the edges with a file or carborundum stone. Repeat for the rest of the set.

House Number Tile

W e shall use a stencil to enamel this house number tile, which is decorated with painting enamels. I used a tile that is available already enameled in white, but before you begin make sure your tile will fit into your kiln. My tile measured approximately 5 × 3 inches, and it fitted diagonally into the kiln I have used for all the projects in this book. You could achieve a similar effect by using two square tiles, each 3 × 3 inches, placing the number centrally across the two. Use a hard white for the enamel background, and follow the instructions for the previous project until the end of step 6. To get a really good finish on your tile, rub down the white enamel with diamond abrasive, a carborundum stone, and grade 240 wet and dry papers until you have a perfectly smooth coat. Refire the tiles to return them to a gloss finish.

You will need

1 oval, pre-enameled tile
Stenciling paper
Pencil
Craft knife or scalpel
Cutting mat
Scissors
Ruler
Set-square
Wet and dry papers
Gum
Paintbrushes
Powdered, sapphire blue enamel –
    washed, dried, and ready for use
Sieve or sifter
Large support
Firing fork
Carborundum stone
Spatula
Red painting enamel powder
Blue painting enamel powder
Lavender oil
Tubes of enamel paint (instead of
    powdered enamels)

1

1 Sketch out the number of your house, or find some suitable numbers for your tile that you can copy. Trace the design, and transfer it to paper to make a stencil.

2

2 Work on a cutting mat or similar, and use a craft knife or scalpel to cut out the numbers to form the stencil. Find and mark the horizontal and vertical center lines on the stencil.

3 Make sure that the tile is clean by rubbing it lightly all over with wet and dry papers. It is easier to mark your pattern on a slightly matte surface.

3

4 Lightly mark the center lines on your tile in pencil, and sketch in the position of the border pattern.

5 Apply a coat of gum to the area where the number will be on the tile, and hold the stencil over the tile or rest it lightly on it.

6 Sift the blue enamel through the stencil onto the tile.

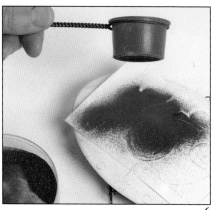

7 Carefully remove the stencil.

8 Touch up the enamel around the numbers, and brush away any excess enamel with a moistened paintbrush. Place the tile on a large support, and lift it carefully into the kiln. The area of enamel at the back of the kiln will fire before the area at the front, so you should turn the tile around so that the enamel fires evenly. Prop the door open, lift out the tile on the support, place it down on the tile immediately in front of the kiln and pick it up the other way round with a firing fork, replacing it in the kiln for further firing.

9 When the blue enamel is smooth, remove the tile from the kiln and allow it to cool. Rub down any unevenness of the numbers, using a carborundum stone under running water.

10 Put small heaps of the painting powders for the decoration at either end of the glass tile. Use a spatula to put a drop or two of lavender oil next to the painting enamels, then gradually pull the oil into the powder, pressing down hard on the tile to mix it in well. The lovely smell will linger for days! The mixture should be a soft, creamy consistency for painting. Do not make it too thin, or it will disappear in the firing; if it is too thick, the finish will be dull.

11 Use a fine paintbrush to paint in the border decoration, one color at a time. Leave to dry on top of the kiln for about an hour, unless you have used water-based tube paints. Before firing, check that the paint is dry by placing the tile at the mouth of the kiln to see if evaporation takes place.

12 Fire until the painted enamels take on a glossy appearance. Turn the tile around if necessary, then remove and cool slowly in front of the kiln.

Enamel can be used in all sorts of different and interesting ways. So far we have been mostly sifting dry enamels onto copper. The next projects involve wet enamel laying. The different enameling techniques have mainly French names, which precisely describe the method used to apply the enamel to the metal, and in this section and the projects that follow we shall put into practice some of these more advanced techniques. The various techniques are described below.

EN BOSSE RONDE

If an enamel is laid directly onto a three-dimensional piece, where it is not contained either by cells or within a recess in the metal, it is

Jay's wing and woodcock wing brooches using champlevé and basse-taille on silver. JANE SHORT.

described as en bosse ronde. The project to make a silver necklace, which involves laying enamel on silver that has been shaped and marked, is a good example of the technique. Other enameling techniques can be used with en bosse ronde – for example, the surface underneath can be engraved or textured to enhance the enamel, and cloisonné wires can, of course, be used to separate and build up a picture or pattern.

CHAMPLEVÉ

The word champlevé means "raised field," and, when it is applied to enameling, it means that the enamel is laid into a depression made in the metal and then built up in layers until it reaches the same height, or is flush with, the surrounding metal. Depressions can be made in the metal for champlevé enameling in several ways. The most effective method is to "chisel" out the area to be enameled with graving tools that are especially shaped for cutting fine lines, and then chiseling out larger areas and neatening up the

Silver enameled bowl with white and 18-carat gold enamel strawberries and leaves. GERALD BENNEY.

inside of the lines. Engraving means that the enamel is not exposed to the vagaries of soldering or the hazards of etching, and the base area can be engraved in such a way as to leave a wonderfully reflective area on which to enamel. However, engraving is a highly skilled technique, which takes a long time to acquire, so to begin with it is better to use other ways to achieve the same end.

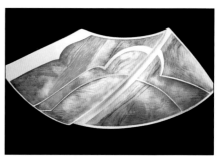

Champlevé enamel, silver, and 18-carat gold rock form. SARAH LETTS.

Soldering

A piece of silver or copper pierced with the desired pattern can be soldered on top of another piece of silver, thus leaving the pierced areas free to fill with enamel. The base piece of silver should be about twice as thick as the piece being soldered on top, which should be the approximate thickness of the intended depth of enamel. I would suggest a thickness of no more than 0.5mm – slightly less, if possible. You can use enameling solder to join the top metal to the bottom, but I have generally found that enameling solder is unnecessary in this sort of application. It requires heating up so near to the melting point of silver, which implies firestain, at least, that I prefer to use hard solder. Remember that whichever solder you use, all traces of solder must be removed from the areas that are to be enameled. If they are not, the heat in the kiln will blacken the solder, thereby turning transparent enamels cloudy and dull.

Champlevé enamel brooch on photoetched silver. GUDDE JANE SKYRME

Etching

Etching is an excellent way of taking out silver and copper to leave areas for enameling. The main problem with this method is that once it starts to bite, the etching medium goes sideways as well as down.

As you will see in the project to make the champlevé pendant, the design is drawn onto the silver and Asphaltum applied to the areas to be protected from the etch. It is essential that the varnish is painted onto clean metal, because it has a tendency to lift away from a dirty or greasy surface when it is immersed in the etching solution.

When you are etching a piece, it is advisable not to cut out the exact shape until after all the etching is completed. This prevents any etching fluid leaking through the edges of the varnish and spoiling the carefully cut edge of your piece. The varnish should be completely dry before it is placed in the etching fluid – if it is still tacky to the touch, it needs more drying. Generally, the slower the etch for enameling, the better the result, because undercut or ragged edges are unsuitable for enamel.

You can buy etching fluid from your enameling supplier, or you can make up a mixture of three parts water to one part nitric acid, or six to eight parts water to one part nitric acid, which gives a longer etch. An etch that is suitable for copper and that may take a couple of days can be made from 7.5 parts water to one part ferric chloride, which should be used with a small air pump, of the kind available from pet shops that stock fish aquaria.

While the piece is being etched, wipe gently across the surface of the exposed metal with a feather. This clears away the metal already etched, and exposes the next layer. You should be able to see bubbles slowly rising to the surface, but if they are too vigorous, the etch is too strong, in which case, carefully pour in one or two more parts of water. You can assess how the etch is going by feeling if there is a noticeable edge with your pointer. Allow an etch to a depth of 0.3–0.4mm before you remove the piece and rinse away the Asphaltum with turpentine. The surface of the silver or copper is then prepared for enameling.

Photoetching

Photoetching is generally only applicable if you are preparing several items that can be etched simultaneously from one piece of metal. It is an extremely accurate method of etching, enabling an exact depth to be achieved, and there are fewer problems associated with undercutting and ragged lines. It is, however, a commercial process, and is not generally possible to do from a small workshop. It is worth finding a local firm that specializes in this process if you think it is applicable to the way you work.

Sterling silver photoetched brooch with transparent enamel and fine silver cloisonné wires. JESSICA TURRELL.

Using castings for champlevé

Most enamelers will tell you not to enamel on castings, which can be full of impurities that cause awful problems when you try to enamel over them. A lot depends on the quality of the castings. Some casting companies will use Britannia silver if requested. This is expensive, but it can be worth it.

Try using castings. They usually have firestain, which should be removed before enameling (see page 30). Any "pickle" or acid remains must be neutralized by boiling the cast pieces in a soda solution, then rinsing well.

If your transparent enamels fire cloudy all the time, despite good preparation, the problem is probably the casting, not the enameling. Rather than struggle with transparents, try using opaque enamels. They may give you better results.

CLOISONNÉ

In this technique, cells or "cloisons" are formed with fine wire, and enamels are laid into these so that they are side by side. The wire used for the cloisons is usually very fine, about 0.3mm in diameter, and it is flattened slightly, either in the rolling mill or by being drawn through a rectangular drawplate. The wires can be used round, but they will appear thicker in the finished piece. The wires stand upright for cloisonné – that is, the flattened sides stand up and the enamel is laid against them. You will find it is easier to include a bend of some sort into a wire, to make sure that it doesn't fall over in the kiln. The wires can then be soldered onto the background metal, but because of the problems solder can cause to enamels, you may obtain better results by first laying a layer of transparent flux and firing it until smooth, and then placing the shaped cloisonné wires

Cast gold units used with plique-à-jour enamel. SARAH LETTS.

Champlevé enamel neckpiece cast in Britannia silver. JINKS McGRATH.

onto the flux, holding them on with a dab of gum, if necessary. Have the picture or template you are recreating in wire close at hand, so that, as you bend each wire, it can be laid on top of the outline to ensure that all the wires will fit as

intended. The wires can be bent with a pair of fine tweezers, small, round-nosed pliers, or flat-nosed pliers, and they can be cut to length with scissors or a small pair of top cutters.

When all the wires are in

position, the piece is refired until the wires just hold into the flux. If the wires sink too deep, the flux tends to creep up the sides, which creates a lighter line when the colored enamel is applied, and this can spoil a piece once it is rubbed down. Any wires that have not been held by the flux can be pushed down with the edge of a pointer. The piece is then ready for the colors to be laid into the cells. Complete one layer at a time between firings, and build up the colors gradually until you reach the top of the wires.

The piece should then be rubbed down with a stone under cold water. This should be followed by the wet and dry papers until the surface is smooth. Any depressions can be refilled and fired, rubbed down, and the piece then given a flash firing. Do not be tempted to start stoning down until the enamel reaches the top of the wires. If you rub down too soon, when the wires are still proud of the enamel, little filings can become embedded in the enamel and are difficult to remove. If you do not want to build up the enamel in lots of layers, use finer cloisonné wires, which will not stand so high and can be filled in two or three firings.

If you need a very thin coat of enamel, you can tack the wires directly to the metal with gum. Transparent enamels can then be laid straight into the cloisons, thereby eliminating the flux coat, but make sure that your colors do not require a flux coat before you attempt this.

ABOVE **Landscape necklace with silver disc, cloisonné wires, and a silver and gold matte finish with coral beads.** JOAN MACKARELL.

BELOW LEFT **Moon earrings with handmade fine silver beads, cloisonné enamel, and crescent moon and full moon stars on reverse.** ALEXANDRA RAPHAEL.

BELOW **Silver enamel brooches with chased silver detail.** SHEILA McDONALD.

1. Silver and yellow enamel goblet.
TAMAR WINTER.

2. Spun silver and enamel bowl.
TAMAR WINTER.

3. Sterling silver peppermill with champlevé and cloisonné.
MAUREEN EDGAR.

4. Table center with "Nautilus" shell. Cloisonné and champlevé enamel with slate base.
MAUREEN EDGAR.

1

2

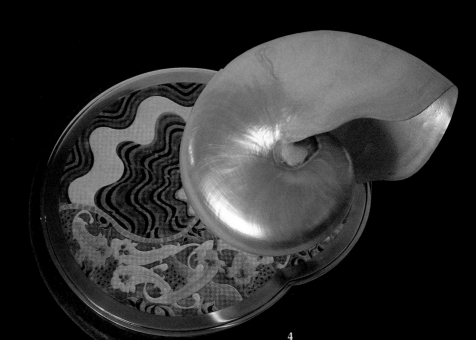

4

5

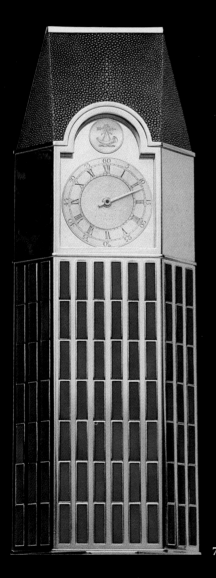

7

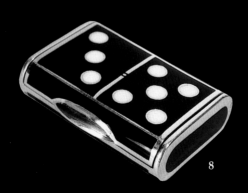

8

5. Goblet.
GERALD BENNEY.

6. Dragonfly clock.
PHIL BARNES.

**7. 24 inch silver gilt clock.
Blue enamel and green
shagreen.**
GERALD BENNEY.

8. Domino box.
PHIL BARNES.

PLIQUE-À-JOUR

Plique-à-jour enameling is like stained glass, but it must have light behind it to show the wonderful colors of the enamels without a metal background. The enamel is held in wire or metal cells, but without the background metal of other techniques.

A shape for plique-à-jour can be pierced out of a single piece of metal, or a shape can be built up using cloisonné wires within a larger frame. There are several ways of laying in plique-à-jour enamel, but there are a few points to consider about the actual structure.

Plique-à-jour Ying and Yang goblets with female and male, water, air, earth, and sun representations. ALEXANDRA RAPHAEL.

First, plique-à-jour does not stay well in sharp corners – curves and rounded ends are much better. In addition, an area more than about ½ inch square will be difficult to fill, and the width of the metal should not be too thin to start with, because stoning and rubbing down can easily result in the piece becoming too thin. Stoning and rubbing should be done very carefully, so that you do not put pressure on the piece. Plique-à-jour enamel is fragile, and cracks easily.

Small pieces can be laid on a clean sheet of mica and, if necessary, held in place with cotter pins. The enamel is then applied as in the project to make earrings with this technique. The mica does tend to lift off and become embedded in the enamel, but there is no need to worry about it until you reach the stoning-down stage, when it gets rubbed away. The final firing of these pieces is done by suspending them in a support, so that no part of the enamel touches the metal support, and flash firing them.

Small pieces can be laid directly on a clean piece of platinum, because enamel does not fuse to platinum, nor does it contaminate it. It is then filled and finished in the same way as laying on mica.

Another way of doing plique-à-jour is to apply the wet enamel straight into the cells without any sort of backing, but using the natural surface tension to hold it in place. Once you have got the knack of dragging the enamel with the water across the cells, you will find it holds quite easily. The piece must then be supported so that nothing touches the enamel while it is fired.

The first firing should take the enamel to the "sugary" look – if it fires for too long, it will pull away to the sides of the cells. The cells are then filled and fired, for just long enough to hold the enamel, until there is enough enamel and it is level with the metal surrounds. Both sides of the enamel are then stoned down and rubbed smooth with wet and dry papers, and finally given a flash firing. If you find that the enamel starts to slump through to the other side during the firing, turn it over for the next firing and take it out of the kiln just before it starts to slump again. Little holes and cracks can be filled up and refired, but if a plique-à-jour piece becomes overworked, you will find that you are curing one problem as the next presents itself. It is very difficult to find a plique-à-jour piece that does not have some little flaw in it.

Larger pieces for plique-à-jour – bowls, goblets, and so on – use a different technique. The piece is made using a straightforward champlevé and/or cloisonné

Plique-à-jour bowl in 24-carat gold using 28 enamel colors. ALEXANDRA RAPHAEL.

technique. The front of the wires and frames are then painted with Asphaltum, and the metal background is etched away to reveal the enamel. This requires a great deal of skill, time, care and attention, but it produces some magnificent pieces.

BASSE-TAILLE

The interest in basse-taille lies in what is behind the enamel. The metal background is engraved or chased, and the pattern can be seen through transparent enamel, which is laid on top. The color of the enamel is affected by the different depths of the lines in the pattern.

Basse-taille brooch and earrings.
JANE SHORT.

The deeper the cut, the darker the lines will appear as the enamel is built up over them. Other patterns can be marked so delicately that they only just show through the enamel. In the project to make an etched and basse-taille pendant, a pattern is etched in the background to show the principle behind basse-taille. Other effects for background patterns can be achieved by using little burr and cutter attachments in a pendant motor. However, the most effective way to produce a good basse-taille is by patiently learning the skills of an engraver.

PAINTING ENAMELS

Painting enamels are applied to work that already has a good coat of opaque enamel. This could be any light color, but is traditionally white. The opaque base coat is gently rubbed down with fine wet and dry papers in preparation for the painting enamels. These enamels are supplied very finely ground, and they are used like paint. To make them usable, the fine powder is mixed with an appropriate medium, such as lavender oil, although some use water. A painting enamel that uses an oil medium will give a slightly stronger color than does a water-based one.

Painted "Poppies" necklace on silver.
GILLIE HOYTE.

Most painted enamel pieces are an intricate build up of different layers and colors of painted enamel. Soft colors, which burn out easily, are usually left until last, while the rest of the picture or image takes shape.

A tiny amount of enamel powder is placed on a clean, smooth glass surface – a 4 × 4-inch glass tile is ideal – and a small quantity of

essential oil is placed near, but not over, the powder. A flat, flexible spatula is used to mix a small amount of the oil with the powder, pressing down to make it into a smooth, workable paste. It should be smooth enough to paint onto the opaque enamel with a fine sable paintbrush.

Before firing, an oil-based painting enamel must be completely dry. These enamels take longer to dry than ordinary enamels, so do not be tempted to fire too soon. Leave the work either close to, or on top of, the kiln for an hour or so, and when the paint is dry it will appear whitish. It can be helped to dry more quickly if you hold the work at the mouth of the kiln and let the oil evaporate. When it goes in the kiln to be fired, watch carefully and take it out as soon as the paint glosses. Painting enamels should be fired at the same temperature used to fire the base coat, but for a shorter time. If painted enamels are overfired they fade rapidly, and it is better to underfire slightly while building up the layers of color, and then to fire correctly on the last firing to obtain the gloss finish. Painted enamel pieces should be counterenameled to prevent unequal tensions building up between the base coat and the painted enamel, which causes cracking in the base coat.

18-carat gold and enamel earrings in sterling silver. TARMAR WINTER.

Painting enamels are usually used to paint miniature pictures or portraits, and they therefore require a certain degree of painting skill. Detail is added as the picture builds up, and highlights are left until the final firing. The enamels are then usually covered with one or two coats of a fine overglaze flux, which acts as a protective layer as well as giving the piece a sense of depth.

Other painting enamels

A less formal approach to painting enamels can be used by applying a single coat of color to add a decoration to your work, as in the project to make the house tile and the painted enamel ring, and this does not need a top flux coat.

Paintbox These enamels come ready-mixed with a medium to form small blocks that fit into a paintbox. They are used by mixing in water with a paintbrush, and then applying the paint to the work. They should be completely dry before firing, and fired until a gloss appears on the surface. These enamels will take on the quality of the enamel beneath them – that is, if they are applied over a transparent enamel, they will appear transparent; over an opaque

enamel, they will become opaque. They will not have such a strong color as powdered painting enamels.

Enamels in tubes Oil- or water-based enamels, also ready-mixed, are available in tubes. It is much cheaper to buy individual tubes than to buy a complete paintbox,

A "paintbox" set of enamels.

ABOVE **Painting enamels in tubes can be either water- or oil-based.**

and oil-based paints will last longer than water-based ones because they do not dry out as readily. On the other hand, water-based paints do not require as long as oil-based ones to dry out on your work, and are fired in the same way.

Grisaille enameling

This enameling technique involves the use of a base coat of black

Grisaille bowl. ALEXANDRA RAPHAEL.

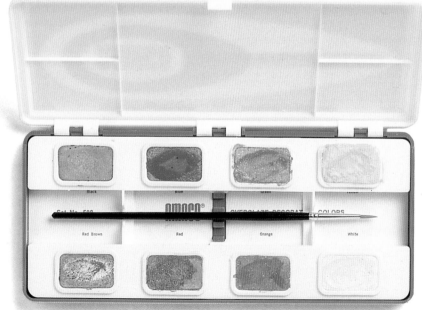

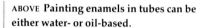

enamel. A very fine layer of white enamel is brushed over the surface, and the design is drawn through the white with a fine point. The work is then fired until the white just begins to gloss. The picture is built up in tones of black and white by subsequent firings, which allow the first white coats to fade into the black and to become gray, while further white coats are built up into highlights. The technique does not seem to be much used these days.

OTHER TECHNIQUES

Lining
Fine lines can be used to enhance outlines or for purely decorative purpose. A simple black line on top of an opaque enamel can be simply drawn with a fine, soft pencil. Rub the enamel with wet and dry papers first, then draw the line and fire the piece at the same temperature as the enamel, but for a shorter time. We have already used this technique in the projects to make the earrings with painted decoration and the house number tile. You can also use an underglaze or overglaze liquid black fine liner.

Japanese sketch. SANDRA McQUEEN.

These come in little bottles, which must be thoroughly shaken before you apply a thin line with a very fine sable paintbrush. Thick lines can look disappointing, so just wipe the liquid away if you do not get it right the first time, and try again.

Underglaze line Paint an underglaze line directly on the metal or on a base flux coat. Allow the glaze to dry before laying on the transparent enamel over the top. Fire the two together. Underglazes come in colors as well as black.

Overglaze line An enamel should be completely finished before an overglaze line is applied in black or metallic luster – that is, the work should have had any necessary rubbing down completed and had its last firing to produce a smooth, shiny surface. The bottle should be thoroughly shaken to ensure that the metallic residues, which naturally sink to the bottom, are well mixed in. Use a very fine sable brush to paint on your design, and allow the liquid to dry completely by holding it at the mouth of the kiln for a few seconds and then leaving it in a warm place on or near to the kiln for at least an hour before firing. A metallic luster fires very quickly – in about 30 seconds – and if it is overfired it will appear dark or get lost in the background color. If it is underfired, it will scratch off with a fingernail. If the luster has been applied too thickly, it will appear as a lifeless blob, so paint as fine a line as possible.

Stitched series. ELIZABETH TURRELL.

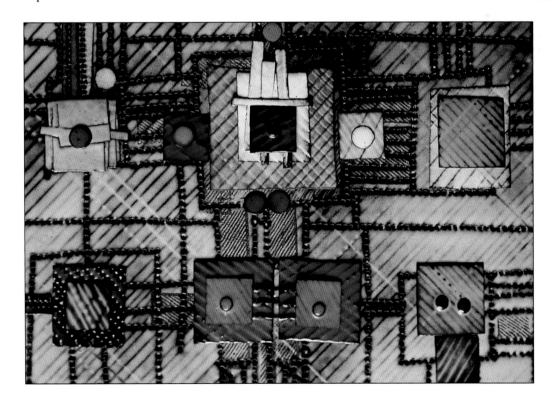

Gallery

1

2

3

4

1. "Desert Dream" copper
sheet with colored areas using
sieved enamel through paper
stencils and silver foil.
STEWART WRIGHT.

2. Enamel and foil panel
with a detail of dual spiral.
SANDRA McQUEEN.

3. Cloisonné and basse-taille
brooches on silver.
JANE SHORT.

4. Silver cloisonné
"Celebration" brooch.
JANE SHORT.

5

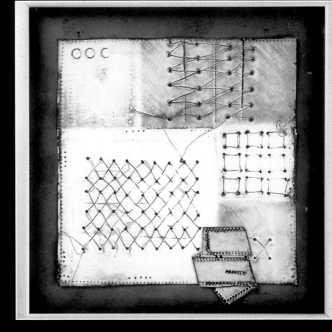

6

7

8

5. Green and blue panel.

Foils

Silver and gold foils are sometimes used as a brilliant background for enamel. Under a transparent color, the foils always have a slightly crinkled look, but they give a wonderfully light and reflective finish to a piece. They need fairly careful use, however. If too large an area is covered with foil, it can look rather garish.

Foils are supplied in very fine leaves, which are kept between two pieces of paper. They are extremely fragile, and should not be touched by your fingers. When you cut foil, draw the shape you need on one side of the paper, and cut around the line with scissors or a craft knife through both pieces of the paper. Fold back the top paper to reveal your foil section. Always lay foil on a flux enamel. Paint a little gum on the flux where you want the foil to sit, and pick up the foil with a fine pair of tweezers or a paintbrush with a dab of gum on the tip. Lay

Pebble boxes with raised forms using a mixture of opaque and transparent enamels. Gold foil matte finish. JOAN MACKARELL.

Sterling silver brooch with cloisonné enamel set with three diamonds. MAUREEN EDGAR.

out the foil as flat as you can, and prick the surface with a very fine pin to allow the air to escape during firing. If you have a largish area of foil to prick, you can make a little tool by breaking the heads off a dozen or so fine pins and pushing the blunt ends into a cork. Check that the points are all level, and use this to prick over the area of foil.

The foil is fired on the enamel until the flux just holds it firm. Transparent enamels can be laid directly on top of the foil, unless you are using any cloisonné wires, when you need to fire a thin coat of flux on top of the foil. Lay the cloisonné wires on the flux, and fire it to fix the wires.

Copper foil is quite different from silver and gold foils, and is dealt with in the next project to make copper foil earrings.

Quantockwood. Copper sheet using sieved enamels through paper stencils. STEWART WRIGHT.

Copper Foil Earrings

This project, which uses copper foil, opens up all sorts of possibilities for you to exercise your creativity. The foil is easy to work with and is not expensive, so you can experiment freely. I have used a liquid flux for the counterenamel and for the flux coat, but you can use other liquid enamels in the same way if you want. When you use liquid enamels, paint them on smoothly, and make sure that they are not too thick. This project uses wet enamels for the first time, so you may find it helpful to refer to the section on page 24.

You will need

Copper foil, about 6 × 3 inches
Pickling solution
Tracing paper
Ballpoint pen
Scissors
Glass brush
Liquid flux enamel
Paintbrush
Tweezers
Mica mat
Mesh stand
Paper towel
Transparent "jaguar" green enamel –
    prepared for wet laying
Transparent may green enamel –
    prepared for wet laying
Transparent pale yellow enamel –
    prepared for wet laying
Firing fork
Small needle file
Hand drill
Jump rings
Earring hooks

Template

1 Anneal, pickle, and clean the copper foil to make it soft and responsive to shaping. Transfer the template to tracing paper, and place the copper foil on a wad of paper or something that will give slightly when you press on it.

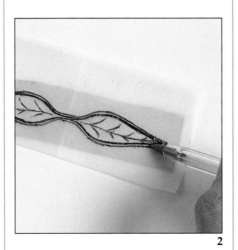

2 Hold the tracing on the copper foil, and follow the outline by pressing down hard with a ballpoint pen – one that doesn't work any more is best.

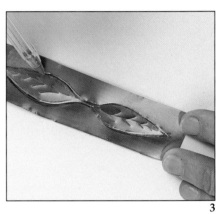

3 Turn over the copper foil, and push the lines through from the other side. This gives an edge and a repousséd look to the foil.

4 Cut out the leaf shape with a pair of scissors.

5 Prepare the copper for enameling by brushing it with a glass brush under the running water.

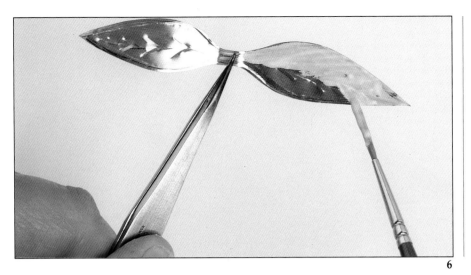

6

8 Wet pack the enamel on the inside of the leaves first. Apply one color at a time, but allow them to merge into each other.

6 Shake the liquid flux enamel thoroughly, and use a paintbrush to apply it evenly to both sides of the leaf. Alternatively, you might find it easier to hold the leaves with a pair of tweezers in the area not being enameled. Bend the leaves slightly so they will stand on the mica mat, then place the mica on the mesh stand and allow the leaves to dry before firing.

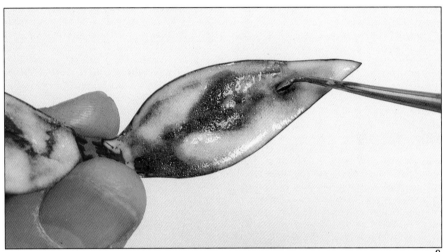

8

7

7 Fire the leaves at over 1,500°F; they will not take long to fire because the copper is so thin. Allow them to cool completely. You will see that where the flux is thinnest, the color of the metal comes through more strongly.

10

9 Remove any excess water by laying the corner of a paper towel against the edge of the enamel.

10 Wet pack the front of the leaves, again applying one color at a time and removing excess water with a paper towel. Allow the enamels to dry out completely before firing.

11 Fire until the colors are glossy.

12 Remove any unwanted enamel in the central area with a small needle file.

13 Bend the leaves over, and drill a small hole through the top of each one for the jump rings.

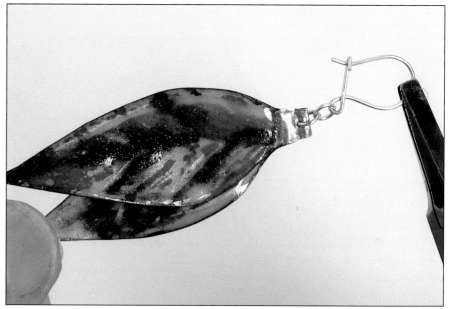

14 Insert a jump ring in each hole, and slip on the ear wire before closing the ring.

Silver Necklace

We now use silver for the first time, so before starting, be sure to read the section on pages 29 and 30 about preparing and using silver for enameling. This is a very simple but colorful little necklace, which is made by laying wet enamel directly onto silver. You will need a piercing saw to cut the silver to shape. If you have not used one before, you will find that they are simple to use and give a very accurate cut. The blade should be tight and springy, and the teeth should face out and downwards. The whole piece is counterenameled with the same colors as on the front, and the highlights on the front are achieved by using the scraffito technique, which we also use for the colored lines in the leaves. I have deliberately left the edges free from enamel as these can be either polished or burnished to give an attractive finish.

You will need

Tracing paper
Pencil
Spray glue or adhesive stick
Silver sheet, approx. 0.5mm thick and
    4 × 4 inches
Hand drill
Piercing saw and blade (size 2/0)
C-clamp
Scrap wood, approx. 4 × 2 × 1 inch
Small needle file
Mica mat
Wet and dry papers
Graver or scriber
Round former (wood or steel)
Glass brush
Small dental tools, homemade tools, or
    quill
Gum (optional)
Pickle solution
Transparent aqua blue enamel –
    prepared for wet laying
Transparent forget-me-not-blue enamel
    – prepared for wet laying
Transparent May green enamel –
    prepared for wet laying
Transparent emerald green enamel –
    prepared for wet laying
Transparent pale green enamel –
    prepared for wet laying
Stand
Firing fork
Carborundum stone
Burnisher
Paper towel
Silver jump rings
Silver chain and catch

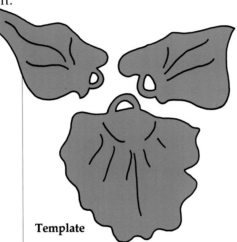

Template

2 Drill holes through the tracing paper, and through the silver where the jump rings will go.

1 Trace the template, and use a spray or stick adhesive to glue the tracing to your sheet of silver.

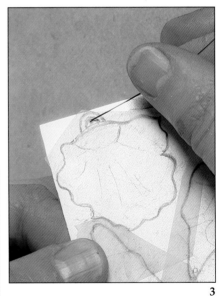

3 Undo the bottom end of the saw blade, and thread the blade through a hole. Fasten it again, and cut out the area inside the line. Undo the blade to release it.

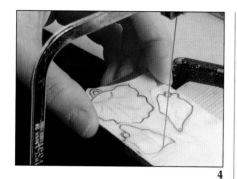

4 Cut out the shapes of the leaves and the flower. You will find it easier to cut out the shapes if you hold a piece of wood to the side of a table with a C-clamp. Use this wood as a "pin" to hold the silver on while you use the piercing saw. You will be able to move the saw around freely without cutting into your table.

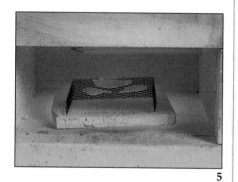

5 Remove the tracing paper and anneal the silver in the kiln, which should be at 1,110–1,200°F. Take the silver from the kiln, quench it while it is still hot, rinse and dry.

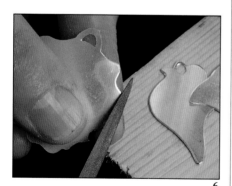

6 File a slight angle at the edges of the flower and leaves, which will give an attractive finish to the necklace, and polish over the file marks with wet and dry papers.

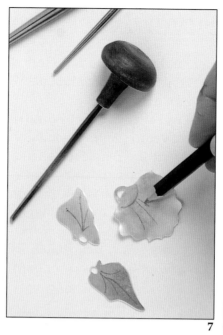

7 Draw in the marks with a pencil, and then use either a graver or the sharp point of a scriber to make the marks.

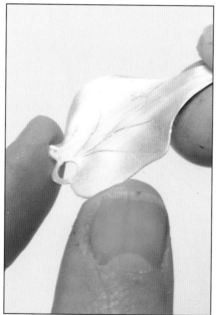

8 After the silver is annealed, it should be soft and workable. Bend it to shape with your thumb and forefinger, or use a metal or wooden former to shape the metal. Clean the bent and marked silver with a glass brush. It should be shiny, and the water should stay in a thin layer all over and not break up to form little globules.

THE FLOWER

1 Apply the counterenamel first. You can paint over the silver with gum if you like, but when you are laying wet enamel this is not usually necessary unless you have a particularly awkward shape. Drag out a small amount of the blue enamel onto your tool or quill, and spread it on the silver. Continue to lay on the blue to within 2mm of the edge, removing any excess water by holding a paper towel to the edge of the enamel. Settle the enamel by gently tapping the side of your work with the tool used for laying the enamel. Allow to dry completely before firing.

2 Fire, remove from the kiln and leave to cool. After firing, the silver on the front will have turned black, unless you have used fine silver. Remove this by immersing the piece in a pickle solution. Rinse well and clean with a glass brush.

3 Lay a thin coat of pale blue enamel on the front. You can hold the flower between your thumb and forefinger or however it feels comfortable, but take care not to touch the enamel.

4 Rest the flower in a stand that will support it at the edges, and fire the first front coat.

5 Remove from the kiln and cool.

6   Put a layer of darker blue on the front of the flower and allow it to dry.

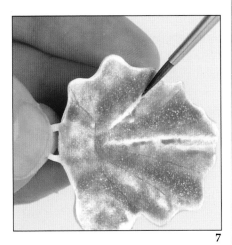

7   Draw a paintbrush or pointer tool through the dry enamel on the high spots of the flower to reveal the paler blue underneath. Brush away any high spots. Replace the work on the stand and fire until the enamel is smooth.

8   When it is cool use a carborundum stone or wet and dry papers to remove any high areas. Always hold the work under running water while you are rubbing down, and because it makes the enamel matte, refire the piece to restore the shine.

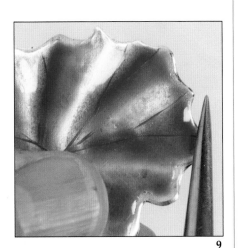

9   Rub a burnisher against the silver. You must support the work carefully while you do this, because pressure in the wrong place could cause the enamels to crack.

THE LEAVES

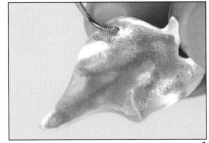

1   Lay the wet enamel on the wrong side first, as you did for the flower. I laid three different greens side by side for the counterenamel.

2   Remove any excess water with a paper towel to help drying, then fire to just past the "orange peel" stage. Remove from the kiln and cool.

3   Pickle and rinse the front of the leaves and clean with a glass brush. Lay the two paler greens on the front of the leaves and allow to dry.

4   Draw fine lines with a pointer through the dry enamel where you want the dark green to be and brush away any high spots.

5   Hold the leaves in the stands and fire until they start to smooth out. Remove and cool.

6   If the exposed silver lines have gone black or oxidized, pickle the leaves until the silver reverts to its normal color, then rinse and dry. Clean with a glass brush.

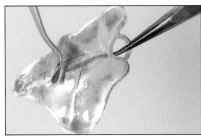

7   Use a fine tool to lay the dark green enamel into the exposed area on the leaf and allow to dry. Support the leaves in the stand for the last firing, place them in the kiln and let the enamel gloss over before removing and leaving to cool.

8   Any oxidization can be removed by pickling, rinsing, and shining with the glass brush. Finish the edges of the leaves by rubbing with a burnisher.

9   Attach the jump ring to hold the leaves and the flower together, then join the leaves to the chain with the remaining jump rings.

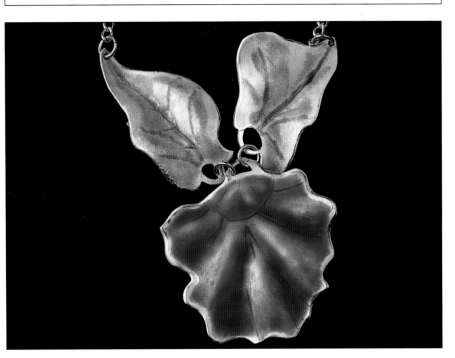

Champlevé Pendant

This project is a silver champlevé enameled pendant. Areas of the silver are etched away to allow enamel to be laid into them and then built up, layer by layer, to the same height as the silver. We will also be laying some cloisonné wires into the etched areas to give fine definition to the separate colors.

If you have a rolling mill, roll the round cloisonné wire so that it has a flat face. If you do not have access to a mill, use the wire as it is – it will just look a little thicker in the finished piece.

You will need

Silver sheet, approx. 1.2mm thick and
   3 × 2½ inches
Dividers
Ruler
Support
Paintbrushes
Asphaltum
Etching solution of 3 parts water to
   1 part nitric acid
Rubber gloves
Feather
Brass tweezers
Turpentine
Piercing saw and blade
Small, flat file
Glass brush
Burnisher
Silver cloisonné wire, 0.2mm or 0.3mm
Wire cutters or scissors
Flat-nosed pliers or tweezers
Gum
Small dental tools, homemade tools, or
   quill
Transparent kingfisher blue enamel –
   prepared for wet laying
Transparent pale green enamel –
   prepared for wet laying
Transparent sky blue enamel – prepared
   for wet laying
Transparent flux (optional)
Firing fork
Carborundum stone
Wet and dry papers (optional)
Small hand drill
Silver jump ring
Silver chain

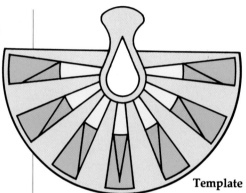

Template

1   It is easier to mark this pattern directly on the silver. Mark a center point with your dividers at least an inch from the top edge of the silver. Intersect the center point with a horizontal and a vertical line. Open out the dividers to 1⅛ inches, and scribe a semicircle from the center point. Open the dividers a further ⅛ inch, and scribe an outer semicircle.

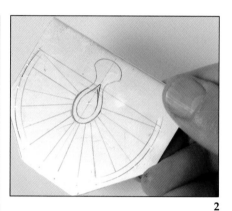

2   Mark the outside of the semicircle at alternating intervals of ¼ inch and ¾ inch, and use a ruler to draw the lines from the center point to the marks on the semicircle. Draw in the shape of the center section.

3   Place the silver on a support so that the edges are free, and carefully paint on the Asphaltum. Be sure to liberally coat all the edges. Leave the front to dry before turning it over to cover the back in Asphaltum as well.

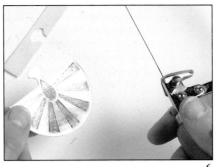

6 Cut out the shape of the piece with your piercing saw, and neaten the edges with a file.

7 Clean up the areas to be enameled with a glass brush, and then polish with your burnisher.

4 Prepare the etching medium, and place the piece gently in it. Always wear protective clothing when you work with acids. Watch the reaction of the etching medium on the silver. It should form small bubbles, which rise gently to the surface. Use a feather to tickle away the liquid from the silver during etching. If it bubbles too rapidly, add a little water to the solution. If you have a strong solution, the etch will probably take about 20 minutes to get to the required depth. A slower etch, which is preferable for enameling, will probably take about three hours.

5 Remove the piece from the etch with brass tweezers, and rinse under the cold running water. The silver should be etched to a depth of 0.2–0.4mm, and you should be able to see and feel a distinct edge. When you have etched to the right depth, clean off the Asphaltum with turpentine.

8 Bend up the cloisonné wires to fit across and diagonally in the areas to be enameled. Make a three-sided shape for the center section, and a tear shape for the center top. Cut the wire with top cutters or scissors, and use small, flat-nosed pliers or tweezers to bend the wire into shape.

9 Put some gum into a section of your palette. Hold the cloisonné wires in your tweezers and dip the wire into the gum. Place the wires into the etched-out areas.

10 Continue gumming and placing the cloisonné wires until they are all in place and sitting flat.

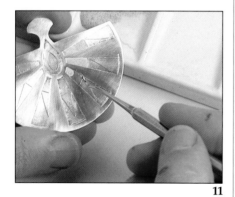

11 Use a small enameling tool to lay the first color in the top sections of the piece. Do not try to fill it all up in one go – it is better to lay a few thin layers than one thick one – and tap the side of the silver to help the enamel settle evenly. Try not to let it dry completely before you add the next color.

12 Carefully lay in the second color.

13 Lay in the third color, and allow it to dry, removing any stray grains of enamel from the surface of silver with a moist paintbrush.

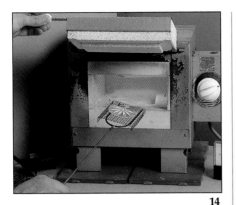

14 Place the piece in the kiln and fire, removing just before the enamel goes smooth. Allow to cool.

15 Gently push down any protruding cloisonné wires with the edge of a burnisher. Refill the cloisons. Allow to dry before refiring.

16 Fire until the enamel is the same height as or just slightly higher than the silver.

17 When the piece is cool, hold it under cold, running water, and use a carborundum stone to stone the enamels down flat. Use wet and dry papers grade 240–600 round the carborundum stone to rub away the marks made by the stone. After stoning the enamel should look uniformly **matte** when it is dry.

18 Refill any hollows with the correct colors, refire the piece, and stone down again until it is flat. If you want to give it a last firing to restore the shine to the enamel, now is the time to do so. I left this piece matte.

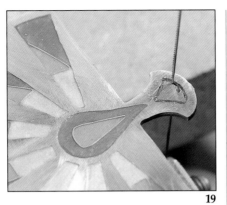

19 Draw on the area that is to be cut away for the jump ring. Drill a hole and thread through the piercing saw blade so that you can cut away inside the lines.

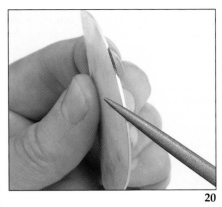

20 Finish the edges of the silver by rubbing firmly with a fine file or a burnisher. Then fix a silver jump ring through the piece, and slip in the silver chain before closing it up.

Basse-taille Pendant

Like the previous project, this pendant uses etching, but it goes a stage farther, etching outlines into an already etched area. This is a technique known as basse-taille, in which the detail is defined under the enamel rather than by cloisonné wires or cells. It is more usual to engrave the pattern for the background, but this requires quite a lot of skill and practice, and, as you will see, etching can produce a similar effect.

You will need

Silver sheet, approx. 1.2mm thick and 2 × 3 inches
Pencil
Stand
Asphaltum
Paintbrush
Etching solution of 3 parts water to 1 part nitric acid
Rubber gloves
Feather
Brass tweezers
Turpentine
Dividers
Piercing saw and blade
Small dental tools, homemade tools, or quill
Transparent rich pink enamel – prepared for wet laying
Transparent sea green enamel – prepared for wet laying
Transparent aqua blue enamel – prepared for wet laying
Transparent rich orange-gold enamel – prepared for wet laying
Transparent sapphire blue enamel – prepared for wet laying
Firing fork
Carborundum stone
Wet and dry papers
Burnishers
Small hand drill with two sizes of bit
Silver jump ring
Silver chain

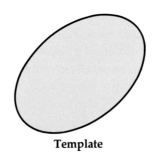

Template

1 Draw an oval about 1½ × 1 inch in the center of your piece of silver. Support the silver on a stand and carefully paint on the Asphaltum up to the oval line you have just marked, making sure that you give the edges a good coat of varnish.

2 Allow the varnish to dry, then turn over the silver and cover the back with Asphaltum. When it is completely dry,

place the silver in the etching fluid, the right way up, and from time to time gently brush the surface of the silver with your feather. Watch carefully to make sure that the acid does not bite too quickly. It tends to undercut lines if it works too fast.

3 When the etch is about 0.2–0.4mm deep, use brass tweezers to remove the piece from the fluid. Remember to wear rubber gloves. Check the quality of the Asphaltum. If it is starting to pull away anywhere, it is better to redo it than to risk putting the piece back in the etching fluid. Remove the Asphaltum with turpentine, and then paint on another coat. You may be able to do localized areas by lifting away the offending pieces, making sure it is completely dry, and then repainting.

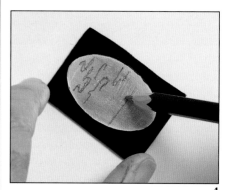

4 Use a pencil to draw in the design that is to be etched further.

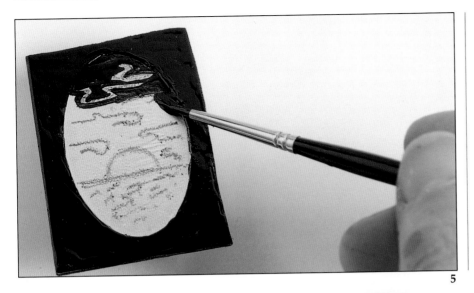

7 Remove it from the acid with brass tweezers when the etch is the correct depth, and remove the varnish with turpentine.

5 Paint the Asphaltum around the pencil lines, making sure the edge of the oval is well protected.

6 Replace the piece in the acid to etch out the lines.

8 Open your dividers to about ⅛ inch and place one point on the inside edge of the oval. Draw around the outside with the other point. Leave a little extra silver at the top of the pendant for the fitting.

9 Use a piercing saw to cut around the outside line you have just marked with your dividers.

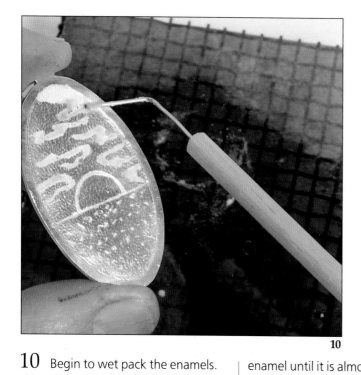

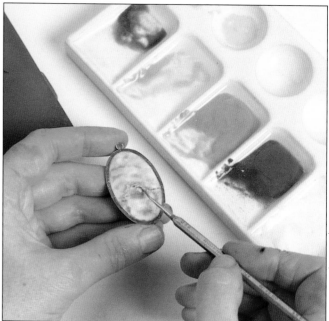

10 Begin to wet pack the enamels. Place the enamel in the lined areas first, and then lay a thin layer on the surface. Allow the pendant to dry thoroughly before firing, then fire the enamel until it is almost smooth. Remove from the kiln and leave to cool down.

11 Lay in the next layer of enamel, mixing the colors by laying one color first and adding the next close up to it. Refire the piece, remove from the kiln, and cool.

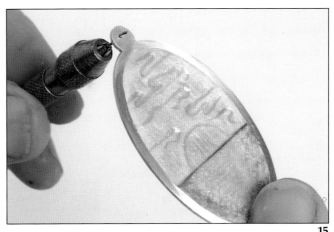

13

15

12 If the enamel is level with, or slightly higher than, the surrounding silver, you can start to stone it down. Use a carborundum stone under running water, then remove the stoning marks with the wet and dry papers wrapped around the stone.

13 Continue until the enamel is smooth and level. Dry it well to see if there are any low spots, and, if so, fill them carefully with the appropriate color. Refire before rubbing down again. When you are satisfied, give the piece a final firing in a slightly hotter kiln, remove, and cool.

14 Polish up the edge of the silver by rubbing with a burnisher.

15 Drill a small hole in the top, and open it out with a larger drill bit or with a piercing saw. Fix the jump ring through, slip the chain through the jump ring, and close the ring.

Enameled Rings

In this project we shall make a napkin ring and two silver rings. The initial stages for both are similar, so the first step-by-step instructions apply to all. The later stages are given separately. You can buy silver tubes in various lengths and diameters.

You will need

Silver tube, approx. 1¼ inches long and 2½ inches in diameter
Silver tubes, ring sizes N and Q
Dividers
Piercing saw and blade
Flat file
Small needle file
Masking tape or similar
Asphaltum
Pencil
Tweezers
Craft knife or scalpel
Etching solution of 3 parts water to 1 part nitric acid
Rubber gloves
Feather
Brass tweezers
Turpentine
Glass brush
Silver cloisonné wire, 0.2 or 0.3mm
Wire cutters or scissors
Transparent flux (optional)
Gum
Paintbrush
Small dental tools, homemade tools, or quill
Opaque white enamel (medium) – prepared for wet laying
Opaque sunflower enamel – prepared for wet laying
Opaque black enamel – prepared for wet laying
Opaque scarlet enamel – prepared for wet laying
Opaque green enamel – prepared for wet laying
Transparent violet enamel – prepared for wet laying
Mesh tray
Firing fork
Spatula
Painting enamels, red and green
Burnisher

RING 1

1 Inscribe a line on the silver tube by holding one point of your dividers at the end of the tube and the other at the length you wish to cut off.

2 Pierce through the line you have made. File the top and bottom of the silver straight, and smooth with a large, flat file.

3 Use a flat needle file to remove the burrs on the edges of the tube.

4 Mark the central area to be divided with your dividers.

5 Cut a strip of masking tape, and wrap it around the tube between the two lines you have marked.

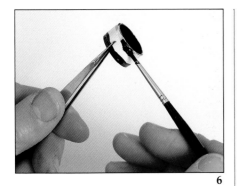

6 Apply Asphaltum to the rest of the tube, including the inside, and remember to coat the edges well. Leave it to dry thoroughly.

7 Remove the masking tape with tweezers, and lift off any stray Asphaltum with a craft knife. Retouch any areas of Asphaltum that may need attention, and leave to dry for a couple of hours.

8 Place the ring in the etching fluid. It should take 2–3 hours to etch, but if it is etching too fast, slowly add a little water to the solution.

9 Draw a pattern on the ring that is to be enameled, outlining the areas you do not want etched away.

10 Paint Asphaltum on these, and leave to dry.

11 Carefully place the ring in the etching solution. Remove both the napkin ring and the ring from the etching fluid when a depth of about 0.3mm is reached.

12 Remove the Asphaltum with turpentine, and wash the ring thoroughly.

13 Use the glass brush to clean and shine the area to be enameled.

14 Use a pointer or fine quill point to lay the enamel into the areas around the pattern. These areas are small enough for you to lay the enamel in without having to use gum, but if you fine the enamel slips, make sure that you remove excess water with a paper towel after you have laid a section, or, failing that, use gum. When the enamel is all dry, place the ring on a mica mat or the mesh stand and into the kiln for firing.

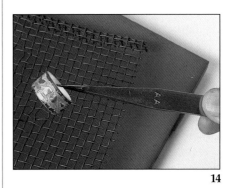

15 Remove and cool. Refill with enamel, dry and refire if necessary. When the enamel is level with the silver surrounds, stone it down under running water with a carborundum stone and wet and dry papers until the enamel is smooth.

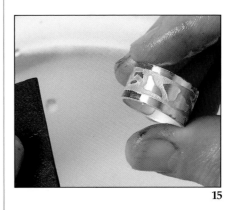

16 Refill any low areas and refire, then give the ring its final firing to restore the shine. It is possible to omit this final firing if you rub down the enamel finely enough with the 600–1200 grade wet and dry papers, which will give a very attractive matt finish.

RING 2

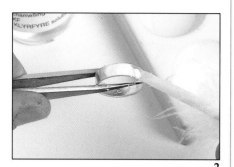

1 Follow the first three steps for Ring 1, then clean the ring with the glass brush and paint a coat of gum all round it.

2 Hold the ring in the tips of the tweezers and carefully lay on the white enamel, drawing away any excess water with the corner of a paper towel as you complete each section.

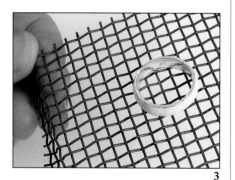

3 Lay the ring on the mesh tray and fire it in the kiln. Remove and cool. If necessary, lay on another coat of white enamel, dry, and refire.

4 When the enamel is cool, rub it down with a carborundum stone until it is perfectly flat and smooth all around, then place in the kiln and refire until the enamel is glossy.

5 Mix the painting enamels and, holding the ring in tweezers and using a very fine sable brush, decorate the ring with one color at a time. Leave the enamel to dry on top of the kiln for about an hour, or help it to dry by holding the ring at the mouth of the kiln until no fumes rise from it.

6 Fire until the colored enamels take on a shine. Remove from the kiln and cool, then carefully file the edges of the ring to expose the silver, burnishing the edges to finish off.

Take care what polish you use on white enamel. Polish can make the enamel dirty and difficult to clean, so try polishing on a test piece first.

NAPKIN RING

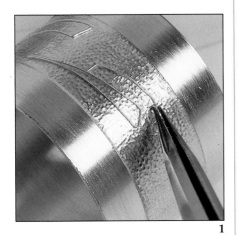
1

1 Follow the first 8 steps for Ring 1. Flatten and bend the cloisonné wire using the wires round if you do not have a rolling mill to flatten them. Hold the wires in tweezers, dip them into the gum, and place them on the napkin ring. Make sure that they fit the curve of the ring. If you find it difficult to get the wires to stay in place, lay a thin layer of flux enamel around the ring. Let it dry well before firing at 1,500–1,560°F, then dip the cloisonné wires into the gum and place them in position on the fired flux. Make sure they are secure, and then fire again until the wires just sink into the flux. Remove from the kiln and cool. Press any stray wires down onto the flux.

2 Lay the colored enamels into the cells using a quill or your fine tool.

3 Lay the yellow around the ring in sections. Paint a small area around the cloisonné wires with gum, and then lay in the color. Draw the water away with the corner of a paper towel. Paint the next section with gum, and lay on the next section of colored enamel. Apply ½ inch at a time.

4 Draw off the water with a paper towel, and continue to lay the enamel in sections until you have completed the circle. Do not worry if it looks uneven at this stage – just try not to put it on too thickly.

4

5 Stand the ring on the mesh tray, and make sure it is completely dry before placing it in the kiln for firing. Remove and cool.

6 Add a second layer of enamel in the same way as you did the first, then fire. Because we are using opaque enamels for this project, a good firing will take longer than for transparent enamels.

7 Rub down the enamel with diamond abrasive and a carborundum stone under running water.

8

8 Dry the ring. Refill any areas that are a little low and therefore still shiny, then refire. Remove from the kiln and cool. Rub down again with wet and dry papers until the enamel is smooth. Give the napkin ring a last flash firing in a hotter kiln and for a shorter time.

9 Carefully clean the oxides from the silver. If you have tested the enamels and found them safe in acid, pickle them to clean them. If not, use fine wet and dry papers grade 600–1200 to rub inside and outside the ring. Do not touch the enamel. On this ring, a burnisher was used to polish the edges, and the silver was left matte.

Box Lid

The enamels used on this box lid are fairly simple so that we can concentrate on using foils and seeing how they react under transparent enamels. I have used copper and a base coat of opaque enamel, with a transparent enamel over the foil. If you prefer to use silver, I would suggest that you apply a transparent or flux base coat, followed by the same or another transparent enamel over the foil.

You will need

Copper or silver sheet, approx. 0.6mm thick and 3¼ × 1 inch
Wet and dry papers
Glass brush
Stand
Gum
Paintbrush
Transparent copper flux (hard)
Powdered opaque sunflower enamel – washed, dried, and ready for use
Sieve or sifter
Firing fork
Carborundum stone
Silver foil
Tracing paper
Pencil
Ruler
Scissors
Cutting board
Spatula
Craft knife or scalpel
Dressmaker's pin
Powdered transparent aqua blue enamel – washed, dried, and ready for use
Distilled water
Small file
Contact adhesive

1

1 Anneal the metal, and clean it thoroughly with wet and dry papers and a glass brush. Support the piece on a stand, and coat it with a layer of gum.

2

2 Sift on an even coat of hard copper flux for the counterenamel.

3

3 Place the piece on the mesh stand, put in the kiln, and fire until the counterenamel just glosses.

4

4 Remove from the kiln and cool. Remove the oxides with wet and dry papers, and clean the surface with a glass brush.

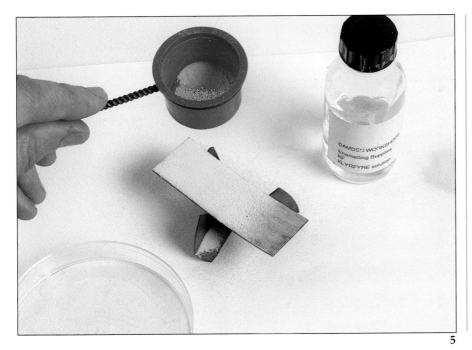

5

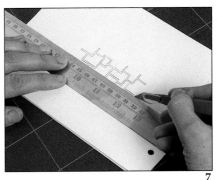

7

7 Cut through the paper/foil/paper sandwich with a craft knife, using a ruler as a guide on the straight lines.

5 Support the piece on the stand, cover with a coat of gum, and sift on the opaque enamel. Carefully remove the piece from the stand, and tip up the stand so that you can clear off the loose enamel. Replace the piece, and fire it until the enamel is glossy and smooth. Remove and cool. If necessary, stone down the surface of the opaque enamel until it is perfectly smooth.

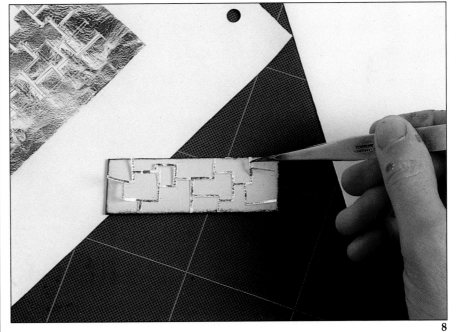

8

6

6 Prepare the silver foil by tracing the pattern onto a sheet of paper that is folded in two, and define the lines with a ruler. Place a sheet of foil between the two sheets of paper and rest it on a cutting board. You can lift the foil from its pack with a spatula or with a paintbrush dipped in a little gum.

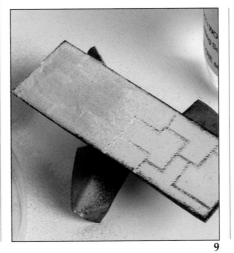

9

8 Paint a thin layer of gum onto the opaque enamel surface, gently lift up the cut-out foil, and place it on the enamel. If you are careful you can adjust the position at this stage. When it is in position, use a fine pin to make tiny pricks all over the foil.

9 Place the piece on the stand and fire it until you see a gloss appear on the surface. Remove and cool. Sift the transparent enamel over your piece, tipping away excess enamel from the stand.

10 Fire the piece, remove it from the kiln, and cool.

11 Add a little distilled water to the transparent enamel, and wet lay some enamel into two or three squares of the pattern. Allow to dry before firing.

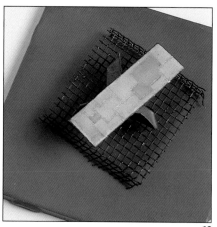

12

12 Remove the piece from the kiln and cool. If the surface is uneven, rub it down with the carborundum stone and wet and dry papers, then refire it. Clean up the copper edges with a small file, and set the piece into the box lid with a contact adhesive.

10

11

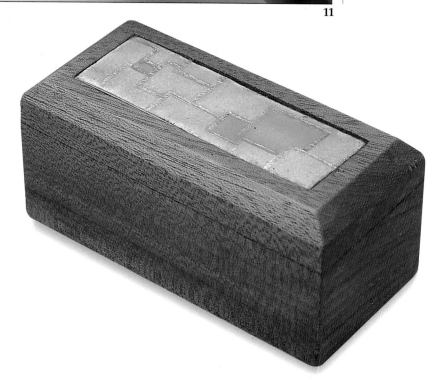

Gold Foil Box Lid

This project uses gold foil under transparent enamel, and you will see how vibrant the colors can be when used over gold. Although I have made a box lid, you could use the same method to make a panel to set into a silver surround for a pendant.

You will need

Circle of copper sheet, approx. 1½ inches in diameter
Stand
Gum
Paintbrush
Copper flux (hard) – washed, dried, and ready for use
Sieve or sifter
Firing fork
Wet and dry papers
Glass brush
Paper
Gold foil
Pencil
Scissors
Dressmaker's pin
Transparent blue flux – prepared for wet laying
Cloisonné wire, 0.2 or 0.3mm
Tweezers
Transparent glossy black enamel – prepared for wet laying
Transparent gold enamel – prepared for wet laying
Transparent rich orange-gold enamel – prepared for wet laying
Transparent orange enamel – prepared for wet laying
Transparent rich pink enamel – prepared for wet laying
Carborundum stone

1

2

1 Anneal and clean the copper circle so that it is ready to enamel. Place it on the stand, and paint on a layer of gum. Sift on a layer of hard copper flux, remembering to tip away the excess flux from the stand. Fire in the kiln, remove, and cool.

2 Remove the oxides that have formed from the other side with wet and dry papers, and use a glass brush to clean the surface. Then place the piece on the stand, paint on a layer of gum, and sift on a layer of hard copper flux. Fire the piece until it is smooth, remove from the kiln, and cool.

3

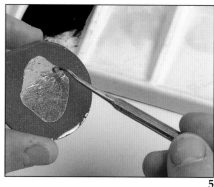

5

5 Place the piece on the stand. Fire it so that the flux holds the gold foil, remove, and cool. Wet pack a fine layer of blue flux over the gold foil. Allow to dry, fire, remove from the kiln, and cool.

3 Place the gold foil between a folded sheet of paper. Draw the oval shape on the front of the paper, and use scissors to cut out the oval through the paper/foil/paper sandwich.

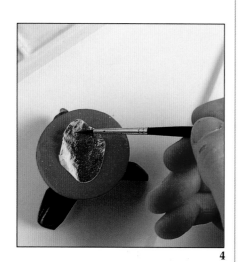

4

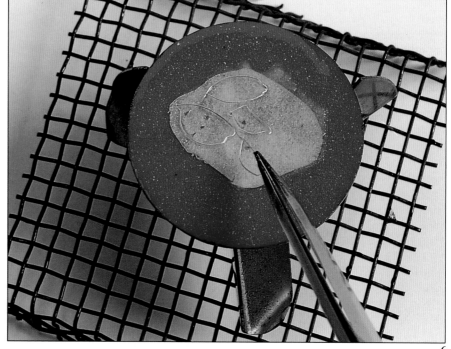

6

4 Paint a little gum onto the fluxed front of the circle, pick up the gold foil with the tip of a paintbrush, and lay it onto the piece. Use a fine pin to make tiny pricks all over the foil.

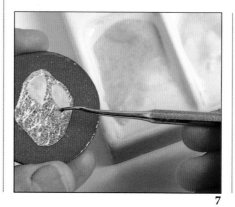

7

6 Bend up the cloisonné wires — using them round if you have no means of flattening them — and place them carefully in position on the foil. Use a little gum to hold them in place, if necessary. Place in the kiln so that the wires are just tacked into the flux as it starts to fuse. Remove and cool.

7 Wet pack the transparent enamels into the cloisons.

8

10

9

8   Carefully wet pack the black enamel between the cloisons, and continue until the black enamel covers the piece. Place the work on the support and dry thoroughly before firing. Remove from the kiln and cool. When using a strong color such as black, you could fire the transparent colors and black alternately. Any strong black enamel is easier to remove from the fired colors than from drying, wet-packed enamels.

9   Continue to wet pack the transparent enamels in the cloisons and to lay in the black in thin layers. Dry, fire, and cool until the top of the cloisons are reached.

10   Stone down the piece under cold running water until it is perfectly level, then rub with the wet and dry papers until it is smooth. Refill any depressions with enamel, and refire. Rub down again if necessary before the final flash firing.

Troubleshooting

ENAMEL CRACKS OR PINGS OFF

POSSIBLE CAUSES

1 Too thick or too uneven a layer of enamel has been laid.
2 Were the enamels fired with a blowtorch?
3 The piece was not counterenameled.
4 The metal is dirty.
5 A setting is creating unequal pressure.
6 The firing temperature of top coat is too low for the base coat.

POSSIBLE CURES

1 Counterenamel and carefully fill in cracks or depressions before refiring.
2 When you fire with a torch, make sure that no part of the flame touches the enamel. Allow to cool without touching.
3 Counterenamel.
4 Make sure the next piece is clean.
5 Try refiring.
6 Refire closer to the temperature needed to fire the base coat.

ENAMEL LOOKS BUBBLY OR HAS HOLES IN IT AFTER FIRING

1 The enamel was too wet when fired.
2 The enamel was too thickly laid on.
3 There is firestain, or impurities are present in the metal – is it a casting?
4 The firing temperature was too high.

5 Layer of enamel was too thin.

1 Refill the holes. Always dry thoroughly before firing. Refire.
2 Stone down gently, rub with wet and dry papers, and refire.
3 Try to avoid firestain – see page 28 for copper and page 30 for silver. If it is a casting, try another one.
4 Rub down with wet and dry papers, and refill the holes. Refire at slightly lower temperature.
5 Clean with glass brush and water. Dry. Add a coat of enamel.

TRANSPARENT ENAMEL LOOKS CLOUDY

1 Badly washed enamel.
2 Firestain in the metal.
3 Possible acid contamination.
4 Traces of solder going black.

5 Temperature of the kiln too low.
6 Enamel layers are too thick.

1 There is no cure. Always wash enamels well.
2 Try to avoid firestain – see page 30 – and use a base flux coat.
3 Always make sure pieces are well rinsed after pickling.
4 Carefully remove the enamel with a burr, and then remove all solder traces. Remember to do this before starting to enamel.
5 Raise the temperature of the kiln and refire.
6 Try a hotter firing. Remember to build up the enamel in several fine layers.

ENAMEL DETERIORATES WITH PROGRESSIVE FIRINGS

1 Enamel that is deteriorating is probably too "soft," and is burning out because of the temperature needed to fire the other enamels.

1 Try lowering the temperature or use only compatible soft-firing enamels. Try organizing the laying of enamels so that you can do any "hard" firing – at high temperature – first. Leave the soft colors until last, or cover soft colors with a clear flux if you need the last firing to be hotter.

ENAMEL HAS BLACK EDGES OR BURNT HOLES

POSSIBLE CAUSES

1 Enamel is too thin at the edges.

2 Copper tends to do this to enamel.

3 Enamel is overfired.

POSSIBLE CURES

1 Clean the edges by pickling and/or wet and dry papers. Relay more enamel, and fire.
2 File, clean, and burnish the copper as near to the enamel as you can.
3 You may be lucky if you refire at a lower temperature. If not, clean thoroughly, refill, and refire at a lower temperature. Check firing temperatures on your test next time.

BLACK SPECKS IN FIRED ENAMELS

1 Dirt in the enamel before firing.

1 Remove the offending specks by carefully drilling or using the diamond burr in a pendant motor. Take care not to mark the metal background. Refill the hole with enamel after rinsing out the hole thoroughly with water and cleaning with a glass brush. Refire.

2 Dirt in the enamel caused during stoning down.

2 Do the same as above. Remember when you are stoning down to hold the piece under running water. Rub with wet and dry papers to remove the marks made by the carborundum stone.

ENAMEL LOOKS UNEVEN

1 The enamel is unevenly laid.
2 The enamel has not been fired for long enough.

1 Refill, fire, stone down, and refire.
2 Fire until the enamel has a smooth surface.

WHITE ENAMEL LOOKS GREEN AT EDGES; REDS AND SOFT ENAMELS HAVE BLACK PATCHES

1 Enamel is overfired or too thin.

1 Put on another layer, if possible, and refire at lower temperature.

OPAQUE ENAMELS LOOK POROUS; TRANSPARENT ENAMELS HAVE A BLOOM

1 The enamels are reacting badly to pickling.

1 Clean thoroughly, lay on another layer, if possible, and refire. Otherwise, after cleaning, refire and see. You could give the piece a matte finish (see page 28), or polish with pumice. Make sure you test for acid sensitivity.

PIECE HAS WARPED

1 Uneven tensions between the enamel and the metal.
2 No counterenamel.
3 The metal is too thin for the amount of enamel.

1, 2 and 3 As it comes out of the kiln, place the piece on a flat, steel surface. Press down on top with a flat iron. Stand bowls upside down on their rims, and hold the iron on top to straighten. This will straighten most pieces out, but the tensions may still be there and cause pinging. Counterenamel if appropriate, or use a thicker piece of metal next time.

YOU'VE DROPPED IT, IT CRACKS, AND BITS FLY EVERYWHERE

Take it and your courage in both hands. Clean it thoroughly under running water, and inspect the damage. If it is very dirty, use an ammonia-based detergent and hot water to clean it, and rinse very well with cold water. Stone down if necessary, use wet and dry papers, then refill the cracks carefully. Give it a flash firing . . . and hope it's your lucky day.

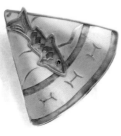

Glossary

Anneal: To soften metal by heating and cooling it at the correct temperatures.

Asphaltum: A bitumen-based compound used to prevent acid eating into metal during etching.

Basse-taille: Enameling technique, using transparent enamels over an engraved, carved, or chased metal surface.

Burnisher: Highly polished stainless steel tool. Hand-held and used to produce a shiny surface when rubbed onto metal.

Champlevé: Enameling technique, using opaque or transparent enamels fired into etched or carved depressions.

Cloisonné: Enameling technique, in which the colors are separated by fine metal wires.

Counterenamel: Enamel fired onto the reverse of a piece to relieve tensions built up between the enamel on the front and the metal surface.

Engraving: Cutting lines into, or removing, areas from metal with a graving tool.

Chasing: Pushing or punching a line onto the front of metal to form a design.

Enameling: Fusing glass to metal at high temperatures.

Etching: Using an acid solution to eat away exposed metal.

Flux: A clear enamel with no metal oxides.

Former: A steel shape used as a support while shaping and forming metal.

Frit: A small "lump" of enamel before it is ground.

Firestain: A blackish shadow that appears on silver as a result of the copper content combining with oxygen during heating.

Glass brush: A brush made of strands of glass fiber, used to clean and shine metal surfaces.

Grisaille: Enameling technique that uses white enamel to build up tones and depth over a black enamel base.

Gum: Material that holds wet or dry enamel onto metal and is safe at high temperatures. Gum tragacanth was once used. The commercially produced gum now used burns out without leaving a mark under the enamel.

Hard enamel: Enamel that fuses at a high temperature, and is usually acid resistant.

High-firing enamels: Enamels that become molten at temperatures of 1,600°F and above.

Kiln: An insulated electric or gas fired furnace used to fire enamels.

Low-firing enamels: Enamels that become molten at temperatures of 1,250°F and below.

Mallet: A wooden or rawhide tool used like a hammer, but which does not mark the metal.

Mica: Mineral sheet that withstands high temperatures, sometimes used as a backing for plique-à-jour enamels. Also used as a clean base for enamels while in the kiln.

Nitric acid: Acid used with water for etching silver and in the removal of firestain.

Oil of lavender: Medium used for mixing fine overglaze enamels.

Opalescent enamels: Enamels that have an "opal" or slightly milky appearance when fired.

Opaque enamels: Enamels that completely cover the metal surface onto which they are laid.

Overglaze colors: Very finely ground colored enamels used for painting, usually onto a previously enameled base.

Oxidation: Discoloration of metal caused by oxygen.

Pickle: A solution, usually sulfuric acid and water, or vinegar and water, used to remove oxides from metal.

Pumice powder: A fine abrasive powder mixed to a paste with water, used to clean metal and also used with "teepol" as a slurry to polish the surface of enamels.

Plique-à-jour: Enameling technique, using transparent enamels suspended in open metal cells, giving a "stained glass window" effect.

Pyrometer: A probe that fits into the back of the kiln, connected to a temperature gauge showing the exact temperature in the kiln.

Quench: To cool metal after heating in water or pickle.

Regulator: A simmerstat with a 1–10 dial, which prevents the kiln from heating further than the selected number.

Scraffito: Technique that makes a line by drawing or scratching through enamel prior to firing to reveal the metal or color beneath.

Soft enamel: Enamel that fuses at a low temperature and is usually sensitive to acid and polish.

Soldering flux: Usually supplied as a white powder, which is mixed to a paste with methylated spirits or water and painted on to silver prior to heating to avoid firestain.

Stoning: Making the surface of enamel smooth and level by grinding with a carborundum stone and water.

Sulfuric acid: Solution used as a pickle with water for precious metals and copper.

Transparent and translucent enamels: Enamels that allow the color of the background metal to be reflected through them after firing.

Useful Addresses for Enamelers' Supplies

UK	US

UK

BULLION: FINDINGS

J Blundell & Sons Limited
199 Wardour Street
London W1V 4JN

Tel: (071) 437 4746
Fax: (071) 734 0273

Cookson Precious Metals Ltd
43 Hatton Garden
London EC1N 8EE

Tel: (071) 269 8102
Fax: (071) 269 8136

Metalor Limited
104–105 Saffron Hill
London EC1N 8HB

Tel: (071) 405 5298
Fax: (071) 405 4844

ENAMELS: KILNS

W G Ball Limited
Anchor Road
Longton
Stoke-on-Trent
Staffs ST3 1JW

Tel: (0782) 313956/312286
Fax: (0782) 598148

Milton Bridge Ceramic Colours
Unit 9, Trent Trading Park
Botteslow Street, Hanley
Stoke on Trent
Staffs ST1 3NA

Tel: (0782) 274229
Fax: (0782) 281591

ENAMELS: KILNS: TOOLS

Gudde Jane Skyrme
Camden Workshops
84 Camden Mews
London NW1 9BX

Tel: (071) 267 4979
Fax: (071) 482 4718

KILNS

Carbolite
Bamford Mill
Bamford
Sheffield S30 2AU

Tel: (0433) 620011
Fax: (0433) 621198

Kilns and Furnaces
Keele Street Works
Tunstall, Stoke-on-Trent
Staffs ST6 5AS

Tel: (0782) 813621
Fax: (0782) 575379

NON-FERROUS METALS

Columbia Metals Limited
Wingfield Mews
Wingfield Street
London SE15 4LH

Tel: (071) 732 1022
Fax: (071) 732 1029

H J Edwards & Sons
93–95 Barr Street
Birmingham B19 3DE

Tel: (021) 554 9041
Fax: (021) 554 7240

Fay's Metals
129 Chiswick High Road
London W4

Tel: (081) 994 4922
Fax: (081) 994 5894

J Smith & Sons Limited
42–56 Tottenham Road
London N1 4BZ

Tel: (071) 253 1277
Fax: (071) 254 9608

TOOLS

Exchange Findings Limited
11–13 Hatton Wall
London EC1N 8HX

Tel: (071) 831 7574
Fax: (071) 430 2028

The Tool Vault
21 St Cross Street
Hatton Garden
London EC1N 8UL

Tel: (071) 430 0577
Fax: (071) 242 7632

TOOLS: CHEMICALS

H S Walsh & Sons Limited
12–16 Clerkenwell Road
London EC1M 5PL

Tel: (071) 253 1174
Fax: (071) 608 1036

US

BULLION: FINDINGS

American Metalcraft Inc.
2074 George Street
Melrose Park
Illinois 60160–1515

Tel: (708) 345 1177/5758

Hoover & Strong
10700 Trade Road
Richmond
Virginia 93236

Tel: (800) 759 9997

Hauser & Miller Company
10950 Lina Valle Drive
St Louis
Missouri 63123

Tel: (314) 487 1311

Myron Toback Inc.
25 West 47th Street
New York City
New York 10036

Tel: (212) 398 8300
(800) 223 7550

ENAMELS: TOOLS

Allcraft Tool & Supply
Company (Mail Order)
60 S Mac Questen Parkway
Mount Vernon
New York 10550

Tel: (914) 667 9100
(800) 645 7124

Allcraft Tool & Supply
Company (Showroom)
45 W 46th Street
New York City
New York 10036

Tel: (212) 840 1860

Amaco Ceramic Supply
4717 W 16th Street
Indianapolis
Indiana 46222–2598

Tel: (317) 244 6871

Bovano of Cheshire
PO Box 250
South Main Street
Cheshire
Connecticut 06410

Tel: (203) 272 3208
(800) 847 3192

Enamel Emporium
113 Meyerland Plaza
Houston
Texas 77071

Tel: (713) 667 6999

Frog Hollow Studio
519 Holloway Avenue
San Francisco
California 94112

Tel: (415) 586 8725

Thompson Enamel
PO Box 310
Newport
Kentucky 41072

Tel: (606) 291 3800
Fax: (606) 291 1849

KILNS

Barnstead/Thermolyne
PO Box 797
2555 Kerpen Boulevard
Dubuque
Iowa 52004

Tel: (319) 556 2241

J M Ney Company
13553 Calimesa Boulevard
Yucaipa
California 92399

Tel: (714) 795 2461

Paragon Industries Inc.
2011 S Town East Boulevard
Mesquite
Texas 75149

Tel: (214) 288 7557
(800) 876 4328

Vcella Kilns Inc.
171 Mace Street Suite B
Chula Vista
California 92010

Tel: (619) 427 2550

Western Kilns Distributors
11990 Avenue Consentido
San Diego
California 92128

Tel: (619) 237 7947

Index